THE BRASS PROPELLER HOOLIGANS

ROY (GRIM) BRISTOW

MINERVA PRESS
MONTREUX   LONDON   WASHINGTON

THE BRASS PROPELLER HOOLIGANS
Copyright © Roy Bristow 1996

All Rights Reserved

No part of this book may be reproduced in any form
by photocopying or by any electronic or mechanical means,
including information storage or retrieval systems,
without permission in writing from both the copyright
owner and the publisher of this book.

ISBN 1 85863 508 X

First published 1996 by
MINERVA PRESS
195 Knightsbridge
London SW7 1RE

Printed in Great Britain by
B. W. D. Ltd., Northolt, Middlesex

THE BRASS PROPELLER HOOLIGANS

To those of us who were not so lucky, and didn't make it this far.

'Finny'	*Johnny*
'Inky'	*Derrick*
'Horace'	*Chas*
'Baron'	*Ken*
'Donga'	*Rod*
'Mousey'	*Jerry*
'Jumbo'	*Ray*
'Zak'	*Fred*
'Wullie'	*Vic*
'Doc'	

'Just dig a hole and pull me through'

Acknowledgements

To 'Ig', 'Waxy', 'Horace' and Robin with my grateful appreciation for their suggestions, corrections of what I had misremembered, and happy reminders of what I had forgotten, as well as to all my Fifty-Fifth Entry comrades for their generous sharing of the experiences.

INTRODUCTION

With the passing out of the One Hundred and Fifty-Fifth entry in the summer of 1993 came the end of the Aircraft Apprentices Scheme, after seventy-four years of continuous, distinguished, and even occasionally faithful service. This is the story of just one of the electricians of the Fifty-Fifth Entry and in some respects, hopefully some of the story of all the apprentices of all trades and all entries.

If I may be permitted, I would like to borrow what I consider to be an apt quotation, from an aged and long-serving Warrant Officer, who remarked bitterly in the sergeants' mess one evening, in the presence of the author, that the Royal Air Force had in fact been cursed since its very inception by two major calamities. One was the noisy self-stropping Rolls Razor and the other the ex-apprentice.

IN THE BEGINNING

It is somewhat distasteful to have to begin all this with a written warning like some label on a bottle of medicine, but in all fairness if you, the reader, are expecting a tale of brave deeds and great exploits in the field of aviation, or perhaps a potted history complete with never before published photographs of the significant developments in the seventy-odd years of the aircraft apprentices era, or how absolutely magnificent we and it all were, and continued to be, then you can put this volume down here and now. My humble intention is simply to describe, as well as my memory serves me, what service as an aircraft apprentice entailed for me and my immediate companions every day, and what it felt like at the time, not what those of a more official turn of mind might like to say that it was, or felt like. Official histories, in my experience, tend to deal in bleak facts, not people, and invariably contain little humour, but most of the memories of that time which I treasure, and have retained, are of young still-emerging personalities in situations from which they still managed, God knows how, to extract some modicum of humour. One of the more sober and responsible personalities from my own entry recently remarked that we really were a bunch of young hooligans in those days, but, even as it was being said, the facial expression softened to one of an all-too-revealing wistful nostalgia. What is quite certain is that absolutely no one came in and emerged at the other end unchanged.

Before I can begin to relate some of my own experiences, I must, of necessity however, explain at least some of the history of earlier significant events which led up to my, shall we say, mutually unfortunate association with the R.A.F., for the benefit of anyone who does not happen to be familiar with this part of Air Force history. For those of us who served as apprentices, it was very much compulsory reading in the first year of service. In 1919, long before I was even born, Lord Trenchard, the Chief of the Air Staff and the true father of the modern Air Force, was still deeply involved in fighting for the very existence of the embryo R.A.F., as always under the cloud of budgetary constraint from its political masters, and with no lack of opposition and continued sniping from both the Navy and the Army, which had gone on since the creation of the R.A.F. in 1918 as an independent service, from the amalgamation of its predecessors the Royal Flying Corps and the Royal Naval Air Service. With all the limitations imposed upon him, Trenchard was still determined above all to be the architect of a force of quality in depth, even if it had to be small, and this was the subject of his paper to Winston Churchill in that year.

Amongst his far-sighted plans was the setting up of short service commissions, the Staff College, the Officer Cadet College at Cranwell, the Auxiliary Air Force as a reserve of weekend fliers, and almost last, but by no means least, the Aircraft Apprentices scheme, the members of which duly became known as 'Trenchard Brats'. He did not quite manage to get it all totally right, however, since, casting his recruiting net even further and wider with an even smaller mesh, he also, most unwisely in my opinion, set up the Boy Entrants system, for those who might have achieved fractionally lower academic standards and be even still further mentally retarded than us apprentices, to provide general low level fodder for the less demanding trades. Thoughtfully it was insisted that they wear chequered hat bands so that they could be more readily distinguished at a distance from us true potential craftsmen. Trenchard had the vision to realise the direction in which aviation was heading, and that it was vital to build within the R.A.F. an ongoing hard core of dedicated skilled craftsmen. Both the Navy and the Army had a long tradition of boys' service going back to the sixteenth century, and he started the R.A.F. along the same well-proven path. Boys would be taken into service when they were between fifteen and seventeen and a half years of age, and trained full-time for three years as craftsmen in a trade in which they would receive full trades union recognition. In my own particular time of service, this was to be extended to include an extra year of improvership to impart some degree of practical experience before going out to serve in the main R.A.F.

In those early times, with its high value on hand skills, this was no small thing since it was the direct equivalent of serving a civilian apprenticeship of at least five years' duration, which in those days often required a premium to be paid by parents and was, what is more, a virtual guarantee of full employment for life, and as a consequence gave an eventual significant rise in community status.

This was, to put it mildly, of considerable interest in the depression years of the thirties which were soon to follow. Thereafter the boys would remain in regular service until the age of thirty, when they would be free to leave the service and take up their trade in civilian life, although still available in times of emergency to be recalled for service. The scheme answered a very real need and, what was equally important, it was cheap. The Fifth entry, I believe, was the first to be trained at Halton which became the Aircraft Apprentices' School.

Without the benefit of any official document giving an insight into his state of mind at that time, I can only suppose that the good lord, visionary that he most certainly was, had fortuitously been spared any panic vision images of the future involving the likely true nature of either myself, or the rest of the Fifty-Fifth entry at the very least. The thinking of that time and that generation, deeply rooted in Victorianism, laid great stress on obedience, discipline and the building of character in the young, who, in the actual event, were quite probably as equally rebellious as any other generation up to the present day. It is not surprising therefore that these ascetic values predominated, and that this tradition tended to be perpetuated throughout the economic hard times of the thirties, and even on into the post-war service.

Through the years there have been times when the entries were large, responding to military expansion periods or a sudden vision of an oncoming national emergency, and, correspondingly, times when the numbers were small, when peace seemed to be well-established, or the civil service was back on track yet again in its seemingly endless search for the holy grail. There have been times as well, when it was not just young Britons who stood at attention on the barrack squares. Many countries have come to realise that the system seemed to work, and was after all preferable to borstal or prison since it was cheaper, and accordingly have sent their own youth to learn the ways. The trades have changed too over the years, responding to and reflecting the changes in technology in the world of aviation. Looking back at the scheme objectively, one must conclude that the R.A.F. did indeed get its hard core of craftsmen and at a very cheap price, particularly in the period between the war years and afterwards. There has also been a tendency for the ex-apprentice to migrate to aircrew after his training, and the ones who did so during the Second

World War made an especially magnificent contribution both to their service and their country. The list of ex-apprentice decoration winners alone bears good testimony to that fact.

Now it seems that the system has finally ended in the light of yet another series of tough budgets. As an ex-apprentice myself, I have experienced my fair share of moments of nostalgia at the thought of this, but, if I tell the story of my memories of what it was like to undergo the experience, then perhaps the reader may come to his or her own conclusions concerning both the value of such experiences to the individual as well as to the R.A.F., and the desirability or otherwise of continuing with it.

Limited Possibilities

Now my story begins to become more personal, and, in all honesty, perhaps it is equally as necessary to tell something of the preceding events in my own life, and the situation in which I found myself, which lead up to my joining the R.A.F. as an aircraft apprentice in the first place, since they provide some rather pertinent pointers to my individual frame of mind as I entered the service, although, in a broad general sense, I was to a very large extent rather the stereotype of others of my generation who followed the same path.

I was born in Grimsby in 1930 into a working class family, the only son of a labourer and his wife. We were not a particularly well off family materially, but no better nor worse than most others who lived around us in those times, although as I grew up I most certainly was never aware of it. I was not a particularly bright pupil at school, at least not in those early years. My type was more the one that was invariably chosen to sing or read poetry at local musical festivals, and whom the parents of daughters of that time simply adored to find sitting next to their own offspring in twin desks in classrooms. Clean, quiet, well dressed and polite, and of course thoroughly boring, but then, as always, it was eventually the daughters who had to come to their own conclusions about the value of that, and what to do about it. I did however eventually manage to overcome my early shyness and get into grammar school on a late developer's scholarship, mostly due to the late influence of an ex-Royal Flying Corps pilot teacher who must have glimpsed some small previously well hidden spark of worth in me, and once there I managed to make decent enough progress.

Like all my generation who grew up in the between-the-wars years, I was fascinated by aviation. The Schneider Trophy events, the long distance and endurance flying records all filled the pages of the press of that time, and our young minds too. I even saw for myself the airship Graf Zeppelin flying over the Humber on its pre-war tour of Britain's shores, and the R.A.F. biplanes lined up outside the hangers at the nearby camp at North Coates, where I would eventually serve my last years in the service. By then it had progressed to being a guided missile station. One could hardly avoid being aware of aviation during the war years with their air raids. Indeed some of my classmates lost their young lives in this way. This was the very visible war with which all civilians could readily identify themselves, rather than newspaper reports of distant events, or even casualty lists which mostly affect the few closely involved. The many bomber and fighter airfields dotted around my native Lincolnshire, both our own and American, gave ample opportunity for plane spotting, even German ones on occasion, and I was able to witness at first hand the technological developments taking place in aviation, such as the rise of the four engine bomber, and the flying bomb. One went to sleep as a child with the sound of bombers gathering overhead for the raids on Germany, and sometimes heard badly missing engines bringing them home in the early hours of the morning. My own interest in aviation went back much further than this, and the names of the new R.A.F. heroes were a mere extension of the ones from all sides in the First World War that I had read about in published memoirs so voraciously as a child; I knew my Richtofen, Immelmann, Boelcke and Udet as well as my Finucane, Bader, Malan and Gibson.

Even so I doubt that I would have come to really consider a career in the armed forces but for a series of dramatic changes in my life that began when I was about twelve years old. After a protracted and distressing illness which continued over the next three years, my mother finally died of cancer at the shockingly early age of forty-five. Although I now realise that most sons tend to hold their mother in rather high esteem, I still, even after all these years, feel that she was a quite exceptional human being, quietly intelligent and far-seeing as well as hard working and devoted. Everything, in fact, that I would have aspired to be myself. During the times that she was in hospital, I had been moved amongst a succession of increasingly reluctant uncles and aunts on a temporary basis, until finally being settled with my paternal grandmother. My mother and I had always enjoyed a very deep and close relationship, and our separation during the illness as well as the end itself had deeply disturbed me, and still does until this very day on occasion. As with most children in the thirties, I had grown up rather remotely from my father, a necessary hard fact of his long working hours and my early bedtime, but, even taking this into consideration, we had never quite managed to develop, in the time that we did spend together, any noticeable degree of that essential closeness, understanding and warm sharing between father and son that is today so very commonplace and noticeable, and which my grandchildren certainly seem to enjoy with their father. Until the beginning of these events, however, I had not even been aware that I had missed out on anything at all. It had all been provided in more than good measure by a quite exceptional mother, but suddenly it was all gone.

My father had continued to live alone at our home during her illness, and, apart from meals which he had at his elder sister's house, to take care of himself during the periods that my mother had been in hospital. During those times, I had seen almost nothing of him, although my grandmother's house and his were well within a ten minute cycle ride or a half hour's walk of each other. By the time that he had finished his day's work, eaten and arrived home, it was already too late for me to go there to him, and he never seemed to find time to come and see how I was managing to get along. My move had also effectively separated me from the companionship of my former playing fellows too. At this point in time my grandmother was old well beyond her years, and ill herself after a lifetime of poverty and overwork, and, whilst I was materially cared for with a roof over my head, clean clothes on my back, well fed and the like, she simply was not able to supply my very real emotional needs of the time.

So from a very happy and carefree early childhood, my life suddenly became rather harsh and lonely, and I had to quickly adapt to taking care of myself with sometimes variable results. My youthful views of what had previously seemed to be a close, supportive and caring family, which I had taken very much for granted, had been dramatically updated within a very short time and I had of necessity learned to trust and rely on no one except myself. I was shocked, deeply angry and cynical and, although I didn't know it at the time, the pain wasn't quite all over yet. The reserved quiet little boy of earlier and better times was not destined to reappear for many years. This does not imply that my father and I were ever in real conflict in any way at that time, or indeed later. It was just that the distance between us now allowed me to see him more clearly as a man rather than a parent, and, what was more, to recognise and understand his limitations, which was of course entirely too early for my still tender years. I was unfortunately well on the way to having a greater in-depth understanding of him than he seemed to have of me, or so I thought at the time.

Within a short time of my mother's death, my father's intention of remarrying became unfortunately, indirectly rather than directly, known to me, and that did seem to be the final blow as far as I was concerned. The possibility of moving back to what was clearly no longer my home was not attractive. Having at least become used to thinking for myself, I also realised that my opportunities in the immediate post-war world of 1946 were rather limited, with the masses of servicemen then being discharged, coming home and expecting to take up their old civilian jobs. Although my personal inclinations were to continue with my education for as long as possible, in view of my father's intentions I was no longer easily prepared to ask him for anything, including keeping me at school for at least two more years, before even attempting to go up to university. Even if I had managed to overcome my reservations about this, it would still have been a financial impossibility for him to go on supporting me, unless I managed to do well enough to gain entry by winning scholarships. With fewer university places available then than now is the case, only a very small top percentage managed to go up at all. This appeared to me to contain a very high element of risk, and to leave me with very little recourse should I eventually prove to be unsuccessful. I would then be caught between being unable to take a place at university, and being too old to be acceptable for any craft apprenticeship either. The alternative was to downgrade my ambitions, and, immediately after taking my matriculation examinations, try and get an apprenticeship at some trade in my home town, which, with its basic economy so heavily dependent on deep sea fishing, was not likely to prove too attractive. That, implicitly, would also leave me with no other possibility but living at home, and yet again being dependent for at least some years. Many of my school friends came to exactly the same economic conclusions about their future prospects, and two who were slightly older than myself were already in the Navy as apprentice artificers, and one was an R.A.F. apprentice.

The better choice, as far as I was concerned, seemed to be to attempt the entrance examinations for the aircraft apprentices' school at Halton. I could get a training, probably in a trade of my own choice with any luck, and also move away from home and at least begin a life of my own. In view of the situation in which I found myself, that seemed no bad thing at all. I had been a member of the

Air Training Corps for a couple of years and had received a decent basic grounding with them. I had attended summer camps at R.A.F. stations, had my first flight in a twin engine Wellington bomber and later ones in that stalwart of Bomber Command in the war years, the four engine Lancaster, and all in all I was reasonably well informed of what service life entailed. I was well aware that the discipline at Halton would probably be very exacting, but I felt that it simply couldn't be as hard as the recent experiences that I had just undergone.

There were a few small obstacles still to be overcome in my intended career path however. One eventually required two operations and a lengthy hospital stay to be able to meet the written medical standards required by the R.A.F., which I had secretly sent for on my own initiative and now closely perused nightly. I didn't have to read very far, and, no matter how I interpreted it, the Air Force pamphlet clearly indicated that in its present state this particular defect would effectively disbar me unless I had some surgery. This presented me with some difficulties. There was no friendly national health scheme in existence in those days, you see, so I would first have to persuade my father to agree to the surgery, since he would have to claim for the expenses on his medical insurance. This in turn implied having to reveal my overall intentions to him sooner rather than later as I had hoped, and, since I was underage again persuading him to agree to my joining the R.A.F. With his basic character being one of extremely cautious reluctance, coupled with an absolute natural flair for coming to a conclusion of no anyway, this was obviously going to be no simple matter, added to which was the fact that he had himself served in the Royal Navy as a boy seaman during World War One, without any great lasting enthusiasm for the armed forces, I might add. So it proved to be, and it was with great difficulty and after many long agonising late night discussions that my father finally agreed to at least accompany me to the family doctor for a medical examination as a preliminary step; and I should perhaps recount what transpired when he presented my body in one hand and the R.A.F.'s written required medical standards in the other at the local surgery one evening, as something of a pointer to the state of working class father-son relationships in the forties. Apart from checking out the state of my blooming general health, the item of main concern was that I was one of those unfortunates who had an undescended testicle which had remained unattended since birth.

When the sound of the summoning buzzer eventually indicated that it was now our turn to enter the doctor's consulting rooms, after the traditional interminable waiting of course, it all started to go unfortunately wrong at once.

"You are not Doctor Pawson!" my father began in a tone of mild surprise followed by a look of deep suspicion.

"No, I am his locum," the young white-coated stranger behind the desk explained with a smile.

Not entirely sure what a locum even was, this only served to further arouse my parent's still at least partially lingering opinion that we had in fact no business being there at all. On the few previous occasions that I had been privileged to observe him in situations of acute insecurity such as these, I had noticed that his invariable response had been to revert to a fall-back position of adopting a face of pure stubborn working class aggression as a defence mechanism, and he did so now.

"I hope that you are not one of them bloody students, because... " he began loudly, but the young doctor, although now driven very much on to the defensive, did finally succeed in calming and reassuring him, at least concerning the validity of his credentials.

Taking his own good time, my father then seated himself on the one chair for the use of patients, removed his cap, placed it deliberately on top of the doctor's papers on the intervening desk, and handed over the pamphlet whilst I stood behind him playing the silent dutiful son.

"It's about this," he offered as an explanation. "He," he said, pointing at me, although I would not have been overly surprised if he had said 'it', since his tone indicated quite clearly his own opinion on the matter, "wants to join."

After the doctor had glanced through the R.A.F.'s written specification for prospective apprentices my father reluctantly came to the deeper purpose of our visit. Personal intimate matters concerning, shall we say, the parts of our bodies which are located below the waistline had never been, and indeed for all of his life, remotely able to be even mentioned between us. Even in late adult life, if I as much as attempted to tell him an even mildly risqué joke in private, he would invariably suddenly change the subject or stop me at once with an admonishing, "Roy!"

It was therefore after much furtive glancing about and in very hushed tones indeed that he eventually brought himself to begin, leaning forward across the desk like an insurance salesman, in a failed effort to prevent my overhearing. "I want you to have a look at his legs," he advised the doctor confidentially.

"Pardon?" the poor doctor asked, also in matching hushed tones, completely baffled. My father, never one noted for his great patience, I must add, became noticeably frustrated by this apparent inability on the part of one who was supposed to be educated to comprehend what to him was obvious. It was now becoming increasingly clear that he might even be obliged to whisper that word which throughout his life was never to be even mentioned between us, although of course these objects of the discussion and the fact that one of them was not located as it should have been was the purpose of the whole visit. I confess to feeling a trifle miffed because, after all, they were my very own.

"His legs, his legs. You know!" my father tried again evasively, but still the doctor did not seem to understand. "Oh hell!" my parent eventually exploded hoarsely in sheer frustration, "I want you to look at his bloody balls, man!"

Even though the cat was now very much out of the scrotum, so to speak, in order to get me even further away out of earshot my father then issued his parental instructions. "Take your trousers off, Roy, and go and lie down on that couch over there." The poor doctor had very little left to say in the matter from then on, but this proved to be the first of many subsequent occasions when I was to be required to remain silent whilst my superiors decided what might best be done with my poor flesh, and so was excellent, if a trifle early, training.

There followed a certain amount of pushing and poking of my unmentionable regions and other places, which my father stood up to cautiously witness over the doctor's shoulder, and a few of the more usual type of measurements and checks followed before I was finally permitted to dress.

"Well! What do you think then?" my father asked impatiently, for all the world as if we had popped in for a building estimate. The by now thoroughly chastened young doctor replied that, apart from you know what, and he would also prefer that I have a more thorough eyesight check by an optician, I appeared to be a very fit young man. My father grunted his grudging approval for what he plainly regarded as a compliment to a piece of his very own work, and we finally emerged with an introductory letter and an appointment for me with a surgeon at the local hospital although, before leaving the room, my father, never one to surrender the initiative once it had been so hard won, had insisted that the envelope remain unsealed and read the contents for himself there and then before it was; as he so aptly put it, "In case I might want a second opinion".

Then my grammar school headmaster, who had enjoyed hopes of my eventually going on to university after completing my matriculation examinations due the following spring, had to be convinced of the somewhat bleak facts of my situation, and the need for me to take some immediate action. After that the rest was up to me, and during and after those stays in hospital I studied hard for the qualification exams which I took in the early autumn of 1946. It did not help that the bias in these examinations tended towards engineering rather than the more academic grammar school subjects which I had been studying until then. I was in hospital for the second of those two visits when my father carried out his intention and remarried, and when I was eventually discharged I came back, at least temporarily, to my original home. Just before Christmas I heard that I was over the first hurdle and was to report to Halton for medical, aptitude and psychological testing in January.

This then was the overall situation at the time of my personal meeting point with Lord Trenchard's earlier plans for the future well-being of the R.A.F. In spite of my very real initial enthusiasm for the R.A.F., this seemingly happy solution to my own personal problems was not as perfect as I certainly believed at the time, as one might guess. Whilst already at least some way down the road of character development that the Air Force intended for all of us, it had until now essentially been my own road, and I had been the one choosing the route and the pace.

I was certainly more mature, independent, and even calculating than my age might have suggested, and more than able to take care of myself after my experiences of the previous three years. This was something that some of my future comrades found rather difficult in the beginning, as I recall. However, on reflection, I cannot pretend that it was a particularly well-balanced young man that the R.A.F. was about to run the rule over. If anything, I had become overly self-reliant, with a profound sense of my own individuality, was deeply cynical of anything resembling authority, and extremely intolerant of small injustices. Paradoxically, and without really realising it at the time, I also had an almost desperate underlying need to belong again, but I did not know where and to what. Coming in the opposite direction on a collision course was the Air Force's well-established intention of eliminating most of that irksome individuality, and moulding all of us into a common something that they considered to be more pliable to their future needs, so a clash of some sort between us was perhaps both unavoidable and inevitable at some point in time, and probably sooner rather than later.

In a very real sense, as well as the story of what it was like to be an apprentice, I realise that this is also a significant part of the story of my own growing up in the early post-war austerity years, through early adolescence and the first full flush of youth towards real manhood, a process by the way that my wife insists is still far from being satisfactorily completed. The two stories are so inexorably interwoven that I feel I would be doing both myself, my comrades and the apprenticeship system a grave injustice, if I did not include at least something of the events of the improvership year, since, in my own case at least, this was a time of a wonderful and continually blossoming sense of utter irresponsibility, and shocking self-indulgence, at least partially due to the nature of the preceding three years, which somehow I was required to equate with the problems of a suddenly emerging career, it seemed. Although most of the essential, still ongoing, character development now tended to occur off duty rather than on, I will do my best to try and include, at least briefly, something of what transpired during the working hours of daylight, in that this provides something of an indicator as to the success or failure of all that which went before, and what followed when we all finally moved to serve in the regular Air Force of those times. However, I will not delay you further with the philosophies of very early middle age, and you will see soon enough how, from what was a most unfortunate beginning, I managed to progress to an even worse middle period. My story begins with the young prince entering from stage left.

'GRIM'

Cats on the roof tops, cats on the tiles
Cat with their...

My reporting instructions, which I reread twenty times at day just to make sure that I had not inadvertently missed something vital in the small print, said that I was to present myself and wait on Marylebone Street Station in London at three o'clock in the afternoon of the fifteenth day in January 1947, where I would join a group of other applicants for transporting to R.A.F. Halton, and where I would then undergo medical, aptitude, and psychological testing. I was to bring only such personal items sufficient to sustain me for five days. I had never had personal items before in my whole life, only toys. No matter how often I read that letter, it never failed to warm my heart, I remember. I recall still how remarkably special and conceited I felt. Can you imagine it - all that testing and just for me. How I would know who my fellow applicants were and how they would recognise me was not explained, however, and that, I admit, left a small shadow over my otherwise general mood of youthful completely unbounded self-confidence. For the first of what proved to be many times, I caught the early morning train from Cleethorpes and enjoyed the parting view of the beaches that I had so often played upon as a child, but now, in my view, I was already very much a man, and that kind of thing was, like the toys, very firmly in my past.

I had never before been to London, or indeed anywhere else for that matter, apart from one day trip as an infant with my parents across the Humber River to the fine city of Hull, which had so impressed me at the time with its vast size that it had remained in my viewpoint as being surely the very centre of the entire universe and the seat of all human understanding. Now faced with London, I accordingly decided not to put my new-found spirit of adventure to an immediate test on the Underground, which was an extremely risky enterprise according to my father, who had apparently tried it once many years before and found himself somewhere near Plymouth. It was, according to him, only for those with a deep understanding of maps and charts, a doctorate in mathematics, or, at the very least, a hardened explorer. Feeling that I did not perhaps as yet quite fall into any of these categories, I instead took a taxi from King's Cross Station when I arrived in the nation's capital, and therefore began my new life in some style, in a manner that I earnestly hoped and fully intended to continue. Thus began the first of many unfortunate trifling errors of judgement and small disasters that were to occur over the next three years as I learned to sail my very own ship, or, should I say, service His Majesty's and later Her Majesty's aircraft. Anyhow, I must have found the one taxi driver in London who did not know where Marylebone Street was, at least when requested in a broad northern accent, of which I must add, I had previously been blissfully unaware, nor did I understand easily his version of English either, so I arrived rather late and more than a little anxious. I had made my first major discovery of the great big world on my very first day: that seemingly all of the English, let alone the Scots, Irish and Welsh do not enjoy the benefit of being united by a common language.

I need not have been too concerned about my lack of punctuality. This was the first of many occasions on which I was to become acquainted with a rather well-known R.A.F. parameter; namely, that if you need them to be there at four o'clock then it is advisable to get them there at the very latest by twelve o'clock just to be sure. Consequently I had plenty of time to become acquainted with some of my future fellow sufferers all sitting around the station, patiently waiting with small cases for those personal items at their feet. They stood out from the normal run of the mill passengers, I remember, like sore thumbs, and the small shadow lifted. All shared this common look of anxious but joyous anticipation, like Christians waiting for their turn in the arena with the lions, which in fact was not too far from the developing truth. I was in Air Training Corps uniform, hoping to score a few extra points at the interviews that way – cunning young fellow that I was – and so was rather obvious, but I remember for example 'Digger' Soames, who was to become the apprentices' band drum major, particularly well, being as open and friendly as he was to remain over the next three years, and Bill Gibson, who would be the sergeant apprentice acting as the parade warrant officer at our passing out parade three years later on, as calm and collected as was his nature. Eventually we were all shepherded on to a train by various anxious N.C.O.s clucking around us like mother hens, and were unloaded in good time on to the main platform of Wendover railway station in Buckinghamshire.

It was an extremely pleasant little country station, I remember, freshly painted and with flower boxes tastefully arranged in prominent places. What I did not appreciate at the time was that, all being well, it was intended by those in power to be some time before I was to be permitted to see the station again, and that by then all forms of artistry, including the garden variety, would probably be wasted on me, but that was all unknown to me at the time as I have said; and I took in absolutely everything around me with bright and deeply interested eyes, seeing deep significance even in the mundane.

Lorries were waiting to take us to the camp, the first and last time that we would be extended that kind of privilege, and before long we were stood en masse on the parade ground of number three Maitland Wing, our names checked against a lengthy list, given a knife, fork, spoon and mug, and allocated rooms in the adjoining barrack blocks. I drew block seven room five, the nearest of the blocks facing on to the parade ground, and made my way up the three flights of stone stairs for what was soon to become a very much repeated journey. Once housed, we were free to choose any bed in the room that took our fancy, each with its blankets, sheets and pillow placed at its head, and an N.C.O., who lived in the small room inside each barrack room which was apparently called a bunk, was there to answer any of our many questions. The wing was newly opened, and, apart from us, was completely unoccupied at that time.

The other apprentices were housed in numbers one and two wing which were quite some way away, and they had been expressly forbidden to make any contact whatsoever with us humble supplicants under pain of death. They were eventually to be distinguishable from us by the colour of their cap bands, red for number one wing and light blue, later changed to dark blue, for number two wing, whilst we were to be the first bearers of orange coloured hat bands. The ones who were eventually selected that is, for all that fine testing still lay ahead of us. Our room N.C.O.s pointed us in the direction of the cookhouse to get our evening meal and told us that we would receive further instructions the following morning.

The R.A.F. was most painstaking in its early care of us. All the N.C.O.s who shepherded us around over the next few days and supervised our rooms and general welfare had been hand-chosen and posted in specially for the event. You can imagine the type: blooming with patience and overflowing with goodwill. They probably went straight back to their monastery in time for evening vespers and an early night of pressing wild flowers in their Bibles, when the last one of us had signed on the dotted line, come to think of it. The sheer quality of the food during the testing period simply defies description, and was never less than excellent. The cooks were dressed in flawless whites and supervised by no less than a warrant officer, who bore a remarkable likeness to the illustrations in my school books of Charles Dickens's Mr Pickwick. He was round, with sparkling eyes peeping at you over the top of small glasses, and, what is more, he possessed an unfailing good humour. After one sumptuous evening meal he arrived at my table with a huge plate piled high with chocolate biscuits, and he literally begged us to take some. If we couldn't quite manage it now, then they were thrust on us, to put in our pockets and, in my particular case, my A.T.C. forage cap, to take back to the barrack room and eat at our leisure later.

After my recent experience of a somewhat spartan wartime diet this was life with a capital 'L', I decided, if you could only manage to just get over those last few hurdles. The deep cunning psychology was already working, you see. Many years later, when wistfully recalling this happy event, I was advised by a fellow ex-Fifty-Fifth entry brat that other aspirants had not been quite so fortunate. Our future Wing Warrant Officer had apparently been getting his hand in at the chocolate biscuits distribution act too, but, with the more usual Air Force frugality, had been allocating just one per plate. At discrete intervals, as the queue line passed him, he had enquired if the recipient would like another one, and if the reply was in the affirmative his humorous response had been to remark, "Greedy little sod, clear off!" He was never a man to neglect the education of the young in my experience. The day after I signed on, Pickwick strangely disappeared, one in and one out so to speak, and, though I waited for the next fourteen years, he never did put in a reappearance, I am sad to say, nor did the chocolate biscuits to which I still happen to be rather partial.

The other applicants around me were from every corner of the British Isles and, as the tests progressed, results became known, and some were eliminated, it became the custom for the survivors to call each other by their town of origin rather than trying to remember if this was Ted number one or George number three or four. It was in this way that for the next three years or so I permanently stopped being known as Roy and became Grimsby, and finally 'Grim', at least as far as my entry was concerned. In common with many who picked up early nicknames, I don't suppose that more than a small few of the Fifty-Fifth ever knew my real Christian name, but most came to be at least vaguely familiar with who 'Grim' was over the next three years.

Others were not so fortunate as I in their new nicknames. In an entry senior to ours, one rather small boy with an effervescent bubbling personality was initially aptly nicknamed 'Twinkle' which then became 'Winkle'. This proved to be too long in practice for everyday use, and so it was shortened to 'Wink'. The next adjustment was far more cynical and vulgar. A Fifty-Fourth electrician, rather short of stature and with hunched shoulders, was rather cruelly named 'Neckless', which of course soon became and remained 'Necklace' for as long as I knew him. Other nicknames that lasted from those early days in my own entry were sometimes positively inspirational. A future airframe fitter with a somewhat mournful appearance and a pessimistic viewpoint of life to match became 'Happy' King, and his companion became 'Waxy' Crane when it was assumed that Crane could just as easily have been spelt Crayon. Fellow electrician Derrick Rushforth somewhat shy about revealing what he felt was not a particularly manly Christian name, was dubbed and remained 'Charlie'. I can remember, among many others, an 'Inky' with barely more than a touch of unseasonable tan, who, incidentally, went on in the fullness of time to be our entries flight sergeant apprentice. There was an 'Ig' because his surname was Noble. A 'Donga' was thus christened because of an unusually large item of personal equipment, which of course never escapes notice for very long among boys, and a 'Bum' for no better reason than he spoke with a slight American accent. A quiet thoughtful Irishman became 'Bogger' Biggs for obvious reasons, and a common cry to be heard in his room was, "Do we want bigger bogs or Bogger Biggs." There was a 'Skinful' because he was of course teetotal, and such a profusion of Smiths that distinguishing nicknames were a vital necessity rather than a humorous diversion, and so electrician Ken, Jack, Dave 1 and Dave 2 were joined by a 'Baron', a 'Peeby', a 'Smudger', a 'Rigor', a 'Curly' and finally a 'Trumpeter'. The latter in the fullness of time changed his name by deed poll, presumably having felt that enough was surely enough. There were even three Berrys, 'Pinky', Johnny and fellow electrician Robin.

Friendships, even unlikely ones, formed very quickly and in the evenings there were visits, to find out how the day had gone for some of one's new comrades housed in other rooms. 'Waxy', a tall and gangling Londoner destined to be a side drummer in the pipe band, paired up with an impish Manchester boy who was almost minute, and would ask those of us who shared his room, with some concern, how things had fared for 'Little Manchester' today.

For the following days we were formed into groups and taken off by the N.C.O.s for the various tests. Some would be having medical tests, others dental tests, eyesight tests, psychological tests, aptitude tests and finally, when all other tests had been completed, personal interviews. At each stage, boys were eliminated, given travel warrants and sent home. Every evening there would be a group of the newly rejected disconsolate in each barrack room, talking together in hushed tones, sharing their common disappointment, and another group of survivors going on to the next stage. It promoted a feeling that we were the lucky ones, and they the unlucky. Over the next few years I often had good reason to reflect that it was probably the other way round, but our psychological tests had been effective enough to reveal to the R.A.F., even at that early stage, that we survivors were just too dumb to realise it. At that time there was an illusion of there being a golden ring, still just barely out of your grasp, which was being offered to you, as an astute salesman might offer a jar of rhino horn powder to an ageing Japanese businessman in the contemporary world. There was a constant reshuffling of rooms to group the survivors into larger units, but by chance I remained in the same one, and in fact stayed there for over a year.

Whilst others were making friends, I made a remarkable and unpleasant discovery about myself that was previously completely unsuspected. One evening in the barrack room after the day's tests, a group of us were standing around, having some discussion about the day's events, and our youthful views on life. My opinion on some quite insignificant matter seemed to be at variance with one of the others, who, incidentally, was eventually accepted for Cranwell. Suddenly, out of nowhere and to my complete surprise, he slapped me hard across the face for no detectable reason at all, which was surprise number one. It was quickly followed by surprise number two, when, without a single moment's conscious thought entering my head, I punched him really hard to the mouth with everything that I had, and he ended up on his back in the middle of the floor. I had never done such a thing before in my life, and in fact had avoided fights and any form of violence like the plague, even as a child. I didn't know what had provoked him to do this and what was worse, I didn't understand how I had reacted so violently. We both shook hands and tried to forget about it all, but I had shocked myself by behaving in this completely out of character fashion. As I mused this over in my bed that night, I realised that somewhere over the previous few years I seemed to have grown very well-defined limits beyond which it was dangerous to push me, and that in that situation I would fight back, regardless of the consequences. It was obvious that I had better be rather careful about controlling myself in future.

THE SHAPE OF THINGS TO COME

He rode up to the manor, he strode up to the hall
"Gor blimey", said the butler, "he has come to..."

When the end of the tests was drawing close, there happened to be a group of us waiting in the room for our friendly 'minder' one morning when the door burst open and in strode a very imposing figure indeed. He was an R.A.F. Regiment Sergeant as immaculately dressed as any guardsman: khaki battledress, blue beret, gaiters and all. You could have shaved on his trouser creases and used his boots for a mirror. He gazed about him in obvious disdain, his lip curling in contempt like a roll of linoleum. His manner was rather brisk and very much to the point and completely at odds with what we had been enjoying as a norm until that time from the other N.C.O.s. who shepherded us about every day. We were all required, it seemed, to move at once if not sooner and clean the ablutions and the general living area. He would lead the way and supervise. As well as an order, this was also another new experience in the use of the English language with which I was soon going to become all too familiar. The vowel sounds were extended between closely pursed lips so that you became yeew and so on, and had the effect of achieving an instant state of attention. Something in our surprise must have alerted him, for he asked with narrowing eyes,
"Erv yew signed on yert?" Our response in the negative prompted him again, "Thern I'll be berk soown," and he departed as briskly and purposefully as he had arrived, banging the door behind him so that it shook in its frame.
We smiled at each other in sheer disbelief at this character. What we did not yet know was that he was very much a man of his word, and every single one of my entry will be in no doubt at all, about to whom I am referring. A new friend, 'Jim', had entered our lives.
The aptitude tests consisted of sitting in a classroom and determining which of the diagrams provided on an answer sheet corresponded with an actual model being manipulated by a class supervisor. The models consisted of a piece of flat cardboard from

which dangled multiple beaded cords with levers protruding from the sides. The inner workings were concealed behind the cardboard, and the supervisor would manipulate a lever or multiple levers, and the cords would move in response. This very soon became extremely complex with several strings and levers being pulled, and many beads moving up or down in response. With little interest in mechanics, a good guess was the best that I could manage with some of these. Additionally I recall that there were electrical circuits which switched lamps on or off according to switch settings. For me, the electrical ones were so much easier anyway, but, with my trade preference very firmly in my mind, I made sure that I gave them my complete and undivided attention.

I had learned to my surprise, from my private pre-medical check in Grimsby courtesy of my father's insurance company, that I was red-green colour blind and therefore not suitable for some trades according to R.A.F. requirements. I was not so certain that my choice of electrician might be one such trade. It seemed quite likely that they wanted to be rather sure that you could at least sort out the different coloured wires in a cable for obvious reasons. This did not deter me in the least and, being a resourceful kind of chap, I had guessed correctly that the R.A.F. would probably use the same colour dot cards for its tests as those on which I had been tested, and I had managed to borrow a set and memorised the lot, with the help of a school friend who had no such infirmity and an optician parent. The R.A.F. was never to discover my little deception.

The psychological tests were the usual one to one question and answer word association games, where the white-coated interviewer asks you to say the first word that comes to your mind when he says a word. He began with 'water.' I replied 'wine,' and his facial expression revealed a slight anxiety; a possible religious hang-up, he obviously thought, or maybe even an early liking for strong drink. He then said 'choir', perhaps trying to steer me back to the religious straight and narrow, to which I replied 'song'. Now even more anxious that he might have inadvertently made an unfortunate choice of word implying homosexual tendencies, he rather over-quickly asked 'male', to which I replied 'women'. Now thoroughly relieved, he ventured to express the opinion that, with interests such as these, it was highly probable that I would eventually make a half-decent electrician. My own observations at the time, which subsequently proved to be quite reliable, were that those who could neither read nor write were best suited to the trade of armourer, whilst those who possessed just one of these two fine attributes were more than adequately equipped to become airframe fitters, if they were bright enough to open the tin of peanuts placed on the desk in front of them, that is. Those who displayed an obvious overly-active interest in brylcreem, hair oil or other lubricants were best suited to be engine fitters, whilst those who could manage to tell the time other than from the angle of the sun were plainly potential instrument-makers. The remainder were most obviously gentlemen and therefore quite likely to succeed in a career in the electrical art.

My own final interview was a masterpiece of pre-planning. The board consisted of all sorts of civilian as well as service representatives. If there had been someone present from the Royal Academy of Dramatic Arts, I would have been starring as Hamlet at the Old Vic the following week. I was in Air Training Corps uniform, and wore the brass arm badge of a trumpeter. I came in, gave my best salute and waited to be stood at ease. They simply loved it, especially my calling the lady member Ma'am when she asked her questions. She was of the classic grey-haired, middle-aged English spinster type and wore a heavy tweed suit complete with cameo brooch and smelled strongly of lavender water, I remember, and therefore was very much at a time in her life to appreciate being addressed by such a grand title.

The apprentices provided the trumpeters for the annual Armistice Day Cenotaph Parade in those days and it is they who played the 'Last Post' which surely everyone has heard since the advent of television. They were justly rather proud of their role. The Squadron Leader chairman noticed my badge and asked if I thought that I was good enough to play with the apprentice trumpeters. I replied not yet, I would have to work hard. He simply glowed with pride. When it came to light that I had voluntarily had those two pieces of surgery just to qualify, he turned to the others and muttered an approving, "I say!"

Even my grandfather's brother's First World War D.S.O. was mentioned and the board chairman finally turned to the others and said, "This is exactly the type of chap that we are looking for, I'm sure you will agree." I think he would have married me if he could.

It was much later that I heard that everyone had received some sort of similar accolade. So it finally came about that one fateful Saturday evening, just about tea time as I recall, in a room in number three wing headquarters with the football results being announced over the radio in an adjoining room – and wouldn't you know it, Grimsby Town had lost yet again – that I was sworn in to faithfully serve His Majesty King George, his heirs and successors, and I became 583602 Aircraft Apprentice Bristow R.F. in the group 1 trade of electrician.

NUMBER THREE WING

My mind was made up for the stage,
At last my ambition I'd gotten

There followed the kitting out, scalped haircuts and the like, which have been more than adequately described by every single creature who has ever had the misfortune to draw on a uniform. We were marched to number one wing stores, still in civilian clothes, and admitted in groups of ten or so. A civilian tailor with quite incredible powers of observation, considering the thickness of his spectacle lenses, took no more than a cursory glance and handed over a standard sized uniform into which you were obviously eventually intended to fit, even if you didn't just now. Occasionally he resorted to drawing rude pictures in bold slashing sweeps on someone's garment in tailor's chalk to relieve the monotony, and bade the new owner to return in one week's time bearing a handed-over-in-exchange document apparently known as a chit, at which later point by the way it was observable that absolutely nothing at all had been changed except that one now needed to rub out the chalk mark.

His enjoyable act was followed by a game in which a long line of stores clerks standing behind an even longer counter filled a kitbag which each of us was obliged to hold open in front of us in growing wonder, all to the accompanying shout of a supervising flight sergeant.

"Shirts blue other airmen, quantity three," he roared. What other airman one wondered? Why could one not have one's own? Were we to begin our noble service in second-hand garments?

"Size?" demanded your personal bag filler of you. If you hadn't yet understood or failed to reply in time you got standard size fourteen. "Collars blue other airmen, quantity six," the shout went on. "Size?" and so on.

There was apparently no such thing as a slim or generous fitting in R.A.F. parlance, we quickly discovered. If you chanced to be a trifle slow in remembering your particular size in something or other, or they did not happen to have your fitting in stock, then sure and certain disaster was very likely to overtake your pocket at a future date although you did not yet know it, because that kitbag was going to be filled anyway and by any means. Perhaps like a birthday card but no accompanying present, it was the feeling that really counted.

Finally we all hastily obliged the last shout of all by signing, "Here, here, and here," at the behest of the clerk who had filled the kitbag, and made our way out dragging the wretched load, to be replaced by the next anxious group waiting outside. We all then staggered back from the stores in number one wing to the barrack blocks of our number three wing home under the weight of it all, with not the faintest idea of what we really had just signed for, nor in many cases what its purpose was. Further examination after dumping it all on one's bed revealed that an item called a housewife was a small pouch containing needles and thread, buttons and the like, but exactly what was the purpose of a cap comforter which appeared to be some form of shrunken knitted scarf remained a complete mystery, as did just how all those webbing belts and buckles and straps fitted together. Meanwhile it apparently all needed stamping, painting, or sewing on to with tape, with your very own service number. When that had all been accomplished, you were quite sure of your service number at least.

While this was going on, you were able to observe the others who had been inducted earlier already at drill on the barrack square in their new uniforms. I was very familiar with the pleasures of drill from the Air Training Corps, and therefore not in any great compelling hurry to sew on my apprentice's brass four-bladed propeller in a surrounding wheel arm badge, and orange three wing cap band, nor to start polishing my new boots to a high gloss, and shining my buttons. I already had some inkling that there would soon be plenty of that. The more enthusiastic, however – 'Weasel' Foster being one that I can recall – just couldn't get out there quickly enough it seemed, and joyful, youthful red faces could be observed from the barrack windows, swinging their arms like demented windmills and banging their new boots on the parade ground, all to the shout of our regiment sergeant friend 'Jim' who had suddenly reappeared as he had promised, accompanied by lots of like-minded friends.

There had been some re-organisation of barrack rooms so that we were now to be grouped by trades, and block 7 room 5 had been allocated as the main home for those of the electrical persuasion. 'Horace' Wallace and 'George' Hanchett, fellow prospective electricians, were fortuitously, like myself, therefore, not required to move. That does not imply, of course, that we were not required to vacate our home on the third floor burdened with all our new kit, make our way down the three flights of stairs and stand with all the others on the parade ground before being allowed to go back to where we had just come. The Air Force simply did not work in that sort of way, we learned. There were in fact a sufficient number of prospective electricians – 'Ig' Noble, Robin Berry, 'Charlie' Rushforth, 'Pussy' Funnel and 'Jock' Clarke – to overflow into a second room located on the floor below ours, which they shared with a similar overspill group of airframe fitters. In the main, the largest groups by far were engine fitters and airframe fitters. The other trades of electrician, instrument-maker and armament fitter were all of roughly the same size, in the region of twenty of each. From this I concluded that the Air Force obviously had more aircraft in urgent need of repair than it had light bulbs, clocks or bombs.

Purely on the basis that they were in the majority, the engine and airframe fitters chose to make the unfortunate and quite erroneous assumption that the rest of us were somewhat lesser creatures, and adopted contemporary Air Force slang to express this viewpoint. Anything which is a worthless trifle wanted by no one in particular, except those who might happen to have some temporary need or just take a fancy to it, is referred to in R.A.F. parlance as being 'gash', a derivative no doubt of the same Royal Navy expression for the rubbish which is thrown over the ship's side. Thus the electricians, instrument-makers and armourers became referred to as 'the gash trades' by their mechanical comrades. 'Waxy' Crane assured me that in later years he once almost caused a strike at British Airways when he used this expression, but obviously we were not quite so sensitive as our civilian counterparts and in fact regarded it in some way as a compliment. All in all, we were over two hundred in number with twenty five prospective electricians. Those selected to become radar and radio fitters were moved to R.A.F. Cranwell in Lincolnshire for their training and we only saw some of them again on rare inter-station sports events. 'Little Manchester' was one of these, so he was soon parted from his new friend 'Waxy', and my fisticuffs adversary was yet another. The electricians of block 7 room 5 settled down slowly to get to know each other.

Number three wing was newly opened and intended to be made up of two squadrons of apprentices, of which we, the Fifty-Fifth entry, were to be just one. Each squadron had its own commanding officer, a disciplinary flight sergeant, and a number of supporting act corporal clerks, plus handy reinforcements, should we prove to be unruly, in the form of the R.A.F. regiment persons, drill and physical training instructors. Plainly having great faith in Egyptian pyramid organisational structures, like a covering umbrella the whole wing had an overall commanding officer and of course his full supporting staff, including a decorated pilot as the adjutant. This was exactly the kind of man to inspire and elicit the admiration of our impressionable young minds, but sadly it soon transpired that his duties and ours did not seem to take us to the same places very frequently, and before long I didn't have the faintest idea of what he even looked like any longer. It seemed that in our youthful quest for the companionship of heroes we were being started at the absolute bottom instead.

There were exceptions however, and by far the most impressive of these in those early days was Warrant Officer Stevenson, the Wing Warrant Officer, who in the fullness of time became privately and very affectionately known to our entry as 'Steve'. He was the ideal Wing Warrant Officer. The kind of man of such patently obvious and completely natural authority that the R.A.F. used to be able to produce from the ranks not infrequently, and sadly which Britain now no longer seems capable of producing at all. In charge of numbers one and two wing were his direct equivalents Warrant Officer Collins and Warrant Officer 'Boggy' Marsh, both similar types.

Rather different from that kind of imagery, however, was Warrant Officer 'Plum' Warner in charge of the station police, of whom there seemed to be a remarkably large number. He was a kind of grey eminence, a Cardinal Richelieu forever lurking in the background, ready to trap the unwary, feared to the point of terror even by his own policemen, let alone the apprentices. 'Steve' however was someone far removed from anything like that, or the Army Sergeant Major bullying type, equally hard when necessary, make no mistake about that, but utterly fair and thoroughly decent, who chose rather to lead by example, and was respected by absolutely everyone. He was a bull-like figure of a man with a black moustache and a habit of stretching his fingers in his leather gloves as if he was doing some preliminary limbering up exercises in anticipation of strangling someone. In time those fingers became a dead giveaway that his patience was almost exhausted, and when they started to curl then it was high time to be long gone if you valued your freedom. He was apparently, we soon discovered, a property-owning man of millionaire proportions if you took him at his literal word. Every single one of us was shouted at some time or other over the next few weeks, and told to remove our miserable wretched presence from MY parade ground, MY N.A.A.F.I. or some other place that seemingly belonged exclusively to him, and often even without the benefit of actually catching sight of from exactly where the shout came. Despite his size, he was an absolute master of concealment and one learned quickly to always proceed in a rather brisk purposeful manner but at the same time with great caution, and absolutely never under any circumstances to risk cutting the very smallest of corners, such as the barrack square on the way to or from meals for example. What is more, if you did chance to catch sight of him, he presented a very intimidating figure indeed, with the obvious muscular ability to back it all up if need be, and he seemed to possess an almost magical ability to be everywhere at once, particularly where mischief was afoot. Beneath this menacing exterior he managed to barely conceal a genuine concern for what he came to regard as also being MY entry. In turn we all came to respect him, because he was always scrupulously fair as well as being hard, a trait not always shared by many of his wing staff, and by almost none of the squadron staff in my experience. We all learned over time to whom to go and talk when you really needed help and knew that the system would refuse you, and examples of this were far too numerous to need expressing further. When, a couple of years or so later, 'Steve' was being pushed out of the service into an unwanted early retirement, he was chanced upon by one of our entry sitting alone on the hillside behind our barrack blocks one weekend afternoon literally in tears because now he would be separated from our entry and not be allowed to see us finally go. He needn't have worried unduly as it happens, because every single one of us took at least a small part of him with us into the big world R.A.F.

Eventually, when everyone had been inducted, we began three months of basic training to fit us into at least a passing resemblance of the occupants of the other wings. They were still forbidden to have any contact with us until we had been sufficiently moulded, but some of the more enterprising of them of course ignored the order, and slipped across in the dark evenings after their day's training. They brought tales of great hardship, harshness and suffering, at which we newcomers all smiled indulgently. They told of things called fatigues, a more severe punishment known as 'jankers' and an extreme known as being 'in the mush.' We fully expected the reign of terror and discipline which is the lot of all who newly join the military for the basic training period. After that, easier times and more attention to our real education would surely follow, and then we could sit back somewhat, we naively believed at the time, but someone unfortunately neglected to inform our mentors that this was the intention.

At our 'B' Squadron level, after a period of having temporary stand-ins in charge of us, we eventually got our very own squadron disciplinary N.C.O. Flight Sergeant Dick Corser, soon to be reduced to plain common or garden sergeant when his acting rank was taken from him for some reason or other, probably for eating an apprentice whilst on duty. That piece of reverse promotion did absolute wonders for his general disposition, I can tell you. He was no friend of mine at all at that time I must add, but he remained with the Fifty-Fifth on and off for most of our stay at Halton, and is still held in great affection by our entry, I believe to his great surprise, and looking back on those early days, certainly to mine. I recall the first time that I saw Dick Corser in close up. We were all formed up in massed ranks outside the barrack block one day, our usual cheerful innocent selves, when he strolled out from his office, stood in front of us, regarded us long and hard and began our long friendly intimate association with a merry quip of his own.

"Why are you leering in that idiotic manner, apprentice?" he demanded of some poor creature in the front rank for openers.

"Sorry Chief," the unfortunate one apologised, using the familiar title used for that rank. That did it.

"Chief! Chief!" he bellowed, rushing forward like an enraged bull for an instant eyeball to eyeball confrontation. "Do you see any feathers or a bow and arrow or something? Well, answer me then, do you?"

He did not seem to be one to overly appreciate informality, and he then expressed the general desire that we all stand to attention and address him as Flight Sergeant in any dealings with him in future. Accordingly the more brave of us persisted with that particular form of address, long after the unfortunate reduction in rank had taken place. Being an intelligent man, the significance was never lost on him.

There was a certain absolute and reliable pattern of behaviour which all of our many disciplinary masters over the years seemed to have in common. They were not people content to sit on their haemorrhoids, pick their teeth and read their comic behind the comfortable obscurity of their closed office door, when there was even the most remote possibility of finding someone wearing an apprentice wheel to attack. What is more, they were always consumed with a quite incredible degree of nasty suspicion and a complete lack of trust. I remember an early day when one or another of them was going through our massed ranks enquiring about names and the last three digits of one's number, whilst peering closely at one's face. This was of course meant to impress us with the distinct possibility of an unpleasant 'I know you' encounter in the immediate near future. He had, as it happened on this occasion, fortuitously encountered several of our many Smiths already, and was about to locate yet another.

"Name?" he demanded. Our apprentice comrade came smartly to attention in the approved manner and replied, "Smith, sergeant." There was a low snigger from our ranks at this apparent ability to hit the Smith jackpot yet again, which was, however, interpreted by our suspicious sergeant, judging by his facial expression, as perhaps being some sort of attempt to deploy facetious humour at his expense. He plainly thought that this particular apprentice might in fact be no Smith at all but a young hooligan of a very different colour, and so he asked for the last three digits of the service number. Unfortunately this particular Smith just happened to be the possessor of the digits 789, which he duly reported, and of course the further coincidence of this particular sequence yet again appealed to the masses and the snigger was even louder. That was sufficient to totally convince our master that he was right, and this was no Smith at all but a mickey-taking young villain chancing his arm.

"A funny bugger eh!" he began and removed a still-protesting Smith family member from our midst by the arm for a remedial session in the cook house cleaning greasy pans, he vainly desperately trying to produce his identity card as proof of his honesty.

Our daily programme in the basic training phase meant drill and physical training in equally obnoxious doses, sandwiched between lectures on everything a young airman should know, aspire to, or not catch under any circumstances, all on an hourly basis between the hours of six-thirty a.m. and nine forty-five p.m. now and forever more to be called 0630 and 2145 hours apparently, for seven days a week. Our friendly local regiment sergeant 'Jim' reappeared in our rooms, as he had prophesied, as well as on the barrack square and he began to personally acquaint us with some of the finer points in the necessary changes to be made in our young characters as well as our appearance. I recall an early morning inspection of our room with every apprentice at rigid attention when 'Slash' Gwilliam in the next bed space to mine seemed to attract his special attention. While 'Slash' fixed his unblinking gaze on some far object, he minutely examined 'Slash's' face from all angles for some time from a distance of about half an inch before enquiring, "Doo yew shave yert apprentees?"

'Slash' was a tender fifteen-year-old at the time and shyly indicated that he did not, to which our friend replied, "Hum! Just bum fluff, eh? Once a week then from now on." He not only remembered the order but, what is more, he regularly checked for himself that it was being carried out.

I myself came very much under his sphere of special interest on one of the very early days that he had us on the parade ground, and he was putting us through a drill manoeuvre known as 'Tallest on the right, shortest on the left in two ranks size.' This was drill gibberish for sort yourselves out in two lines so that you are standing next to someone of the same approximate height. Of course one first has to find such a creature, and it can take a little time when you are strangers, and some general milling about was evidently quite necessary. Not much later I would become all too familiar with the heights of all two hundred of us and, what is more, to the very inch, even when blindfolded. He bade us all to be quiet in his firm manner, which was the usual shrill falsetto scream, and for the greater part we indeed were able to oblige him, but on this early day I chanced to bump into someone in the general mass movement, and of course, as a well brought up sort of chap involuntarily muttered my apology.

At that point the villain suddenly entered from stage right. The sergeant's face and mine were suddenly only inches apart. He was experiencing some terrible breathing difficulties, I noticed, and was apparently in the early throes of having a fit of some kind, purple-faced and bulge-eyed. The veins of his neck were terribly distorted like ship's ropes. I hoped to ease his obvious discomfort as best I could, and smiled at him. It did not seem to help at all and in fact seemed to make matters worse because he then became extremely abusive. He expressed his serious opinion for all to hear that I had apparently been inducted in spite of some very serious hearing deficiency which he was now personally going to correct, and his whole conversation came out at a hoarse bellow from extremely close range, spit and all. I dropped the smile. The overall trend of all this, I detected, being an intelligent sort of chap, was that he seemed to prefer that I was silent and not only for just now, but into the remote far future. Politeness and smiling were seemingly very much unwanted baggage from now on, and indeed it proved to be so.

'Jim' did not come alone now on his visits, and one of his many accompanying friends was Flight Sergeant 'Jimmy' James, also a member of the R.A.F. Regiment and equally guardsman-like. Despite his strict insistence on our own standards of drill ability, he completely disregarded them himself in practice, we soon observed. When in charge of a parade which required that he march forward alone, halt and salute a commanding officer before reporting the readiness of the parade, he always marched forward all right in the approved service manner, arms swinging to the horizontal and beyond, but always at the last possible moment he would abruptly change his angle of approach to the extremely oblique for some strange reason or other, like a car that has suddenly lost one wheel. He would then come to a hob-nailed crashing, skidding halt. I have seen more than one officer noticeably flinch, believing that he was about to be involved in a major head-on collision. The following salute was with the tips of his fingers approaching the vertical somewhere over the region of his right ear and resembled more a rather crimped Nazi Heil than any R.A.F. salute that he ever taught us, but none of us ever dared to imitate it.

Up in the morning exceedingly early and about our training, we would hear but not see the rest of the apprentices in the other two wings being marched to work every morning by the military and pipe bands. Their route from numbers one and two wings to schools and workshops took them past main point, and well away from our wing area. It wasn't long before those of us with a musical bent began to make their way to the band hut located behind number one wing to enlist in the band of their choice. 'Tex' Ward and 'Clutch' Harris began their musical careers as pipers and this continued long after their service lives in this way, as did Geoff 'Gabby' Garbelt and Roger Barrett amongst others. Alfie Swaine and 'Waxy' Crane started on the route to becoming the band's leading and second side drummers, 'Spike' Kelley and Alf Collier to being a quite classic duo of tenor drummers, and Tom Atkinson, 'Trumpeter' Smith, 'Butch' Stalker and 'Titch' Wilding to becoming trumpeters. I dropped my own intention of joining the trumpeters when I noticed, with a daily rapidly developing acumen, that they seemed to be required to be up and about even earlier than the rest of us in the morning playing reveille, and rather late to their bed at night after playing lights out, whilst in between they had to do everything else that the rest of us did, plus attend to their instrument duties which seemed to again require vast quantities of white blanco and much metal polish. I came to the conclusion that, on balance, the discomforts obviously outweighed the joys, and the crotchets and quavers business was pushed into the farther crevices of my mind, and the trumpet mouth-piece that I had brought with me was consigned to the drawer of my locker waiting for more auspicious times, which in the event never quite transpired.

The daily routine for apprentices is governed by trumpet calls, we soon learned. I can still hum the calls for march on, markers and particularly defaulters, in a kind of conditioned reflex action after more than forty years. One of our entry, 'Jock' La Haye, was already a skilled piper and he was taken into the pipe band immediately. When the rest of us were beginning our running and marching everywhere, Jock was with the pipe band in London at the Lord Mayor's procession. He went on to be pipe major. Others, less talented, were required to serve another long musical apprenticeship before being allowed to play with the band. The interest and enthusiasm was such, however, that number three wing formed its own pipe band within a year and it became a focal point which was at the very centre of our existence as an entry. Anyone with any sort of association with pipe bands will know what I am talking about. When you have marched behind a band playing 'The Black Bear', a traditional end of the working day tune used in many Scottish regiments, and joined in the combined shouted cheer, then you know something about the influence that the pipes have had on all kinds of servicemen over the centuries and the apprentices in particular. Sometimes now when I am part of the audience at military tattoos and a pipe band plays that tune, I will hear another voice out there also make that shout at the appropriate point, and know that there is another ex-apprentice out there.

CLEANLINESS IS NEXT TO GODLINESS

I'm the man who cleans the brass,
I'm the man who wipes your basins when they're dirty

We were not left to enjoy the pleasures of number three wing alone for very long when we were joined by the Fiftieth Entry, who moved over from number two wing, changed their blue hat bands to orange and formed the second 'A' Squadron, distinguishable from ourselves by differently coloured discs behind the cap badge. Their destiny now was to be our minders. They now provided the N.C.O. apprentices who took over the duties and responsibilities of the recently departed monks in our rooms. At this point in time, kindness and simple courtesy disappeared with them forever as they descended the stairs on their way back to the monastery. The Fiftieth were very impressive, I can tell you, and in my view probably the most outstanding entry during my time at Halton. They took themselves and their position as wing senior entry very seriously indeed and life darkened even further, if that was in fact possible.

The corporal apprentice at first appointed to be in charge of my room – joy of joys – happened to be none other than my old former school friend from back home, Jack Gough, whom I greeted of course with my usual friendly, 'What ho! Jack.' Only it wasn't quite good old Jack any more it seemed. He immediately assumed an air that I had just done something rather unpleasant on his boot, and that we had never previously set eyes on one another, and I was not slow to take the hint. He gathered us all round a bed and demonstrated once, and only once, just exactly how you laid out your kit for daily inspection, and then allocated everyone their responsibilities for the many cleaning tasks that had to be done every day. These were by no means inconsiderable, nor entirely concerned with cleanliness either. Apart from our own bed spaces, our room furniture consisted of a wooden table top which was supported on two cast metal black frames, and two wooden benches. The wooden surfaces of all had to be scraped to a shining white with razor blades and then bleached.

The two brooms also issued required the same treatment on their handles, and the metal bucket required burnishing to a mirrorlike surface. Until a satisfactory standard in appearance had been reached which could be sustained by one individual, these required our joint attention just for openers. Jack was and remained absolutely impartial towards me, and I expected and received absolutely no favours. Even on the same train going on leave, I was expected, as a humble 'rook', to keep my distance from both him and 'Smoky' Betteridge from the same entry. Once home and now actually on leave at the same time, he would go back to being good old Jack, my street mate, but at Halton we instantly reverted to the more formal pattern.

Privacy, if there was such a thing, consisted of your own bed space of eight feet by four feet in a barrack room for approximately twenty, a metal one-shelved locker on the wall above your bed with a drawer mounted under it, a wooden foot locker at the end of your bed and a bedside table. The drawer was for personal items like a picture of Mummy and Daddy from happier times and perhaps a pen, and the wooden locker had enough space left over for perhaps one or maybe two small parcel-sized objects, and even that was further restricted when we eventually began our technical education and were issued with our own set of technical manuals. All civilian clothes and my A.T.C. uniform were wrapped in brown paper, tied with string and dispatched home at once as soon as we had been inducted. Only one apprentice to my knowledge retained any civilian clothes on the camp and that was 'Bogger' Biggs, who needed them when he returned home to Eire on leave, but even then they were held in store for him and only issued at the appropriate time. As the years passed, 'Bogger' grew, but sadly his civilian clothes did not and, since he was the possessor of a pronounced native stubborn streak, he declined to purchase new ones since in his view it was the R.A.F.'s responsibility to clothe him now. In the final year he went home looking for all the world like the painting of 'The Childe Harold' with trousers that barely covered his knees and sleeves that ended at his forearms.

Everything else remaining had its own place in permanent kit layouts which were somewhat rather more than neatly folded shirts and underpants. Every item had to be of a fixed size and in its fixed place. The rope of one's kit bag had to be tied with a certain precise number of turns for example, and it was necessary to obtain and use all sorts of formers to hold items in exact and precise size and shape.

The varnished wood of shoe brushes, hair brushes, and even shaving brushes had to be scraped with razor blades to a smooth shining white, and laid out in precise position. One wondered, of course, why on earth they had bothered to go to the trouble of painting them in the first place, but one already knew to keep such Bolshevik thoughts to oneself. All this wasn't just for periodic kit inspections, which were much bigger and far grander affairs, but for every single day.

The more astute soon learned to construct the necessary metal formers from any scrap materials they could lay their hands on, to keep everything in a measurably exact size and shape. The less astute or lazier soon found themselves on punishment. Beds were made up every day with the three mattress biscuits at the head of the bed and the blankets and sheets placed in exactly-sized alternate folds on top, all wrapped up so tightly in one blanket that you could bounce a coin on it. A spotlessly clean towel was laid across the centre of the bed, and the bedside table supported your knife, fork, spoon and mug. If you had the opportunity to lie down or sit during the day, which was an extremely rare event, I might add, then it was on bare bedsprings and only after carefully removing the towel.

One's backpack and webbing equipment were fully assembled into a complete shape, and it was necessary to obtain some former to hold the whole mass in a rigid square form suspended beneath the wall cupboard. Meccano sets that had survived from childhood days were suddenly remembered, sent for, and found new purpose in daily lives. The regiment sergeant was not satisfied with the shape of my pack former, constructed from the only materials available to me, discovered by scrounging from the dustbins at the back of the N.A.A.F.I. canteen. He strongly advised me to write to my father and get him to obtain plywood and metal stringers and screws, and dispatch them post-haste for me to build a better former. I was obliged to smile at the ludicrous thought of my father bothering to do anything like that for me, and I told him so. While permitting himself to be most untypically sympathetic with my home circumstances for the most brief of moments, he was, however, not interested in any excuses, and he made it very clear that it was my problem to conform or else.

My entry was fortunate in being the first to have battledress issued as its working uniform instead of two sets of best blue. Its great virtue was that it had no brass buttons to polish, but unfortunately it soon required much more pressing in order to satisfy the exacting standards of our new masters. I used to wear out a set of buttons through to holes on my dress uniform about every year or so on average, I later found. The boots required nightly work with a spit and polish routine, and one tried one's very best to take care that they never became scratched, whilst at the same time everyone in authority seemed to try to make quite sure that they did. In my improvership year at St Athan, one of the ex-Fifty-Fourth entry, Kim Gooch, with whom I was to serve overseas, used to rent out his best boots to those who wished to be selected as the best dressed on guard parade, and therefore excused the duty, on a no selection no fee basis. Even if they required stuffing with toilet roll to make them fit, the boots alone were widely considered to be a virtual guarantee of being chosen, but any one of us ex-apprentices had equally impressive footwear held back should we ever need it again. Indeed, for the rest of my fourteen years' service, I kept one special pair of boots for just that purpose, and used shoes for normal wear.

Even the question of laundry involved a certain amount of careful thought. One was permitted to send six items each week free of charge. Rarely all of it, mostly some of it, and even occasionally your very own came back. With a basic minimum requirement for one shirt, one towel, one pair of socks, one pair of underpants, one vest, two collars and one set of pyjamas per week, you were already over the limit, so a spot of hand washing also came into your list of jobs to be done.

My friend Jack was soon moved to another room, and succeeded by Sergeant Apprentice Roger Swire in the outer room, Corporal Apprentice 'Akker' Bint occupying the separate inner bunk room, and Leading Apprentice 'Danny' Kay, who slept in the main barrack room with us. There was an initial chilly distance maintained between them and us and indeed with the whole of the Fiftieth that we saw going about their normal training. They were not terribly concerned with becoming our friends, it seemed, and regarded us with a scornful disdain better associated with pond-dwelling creatures or younger brothers. To them we were 'rooks' and far removed in the pecking order.

They really were a splendid entry, I can still reflect. Many of them, like 'Straw' Hall and 'Mick' Strickland for example, were already well over six feet tall and splendidly fit, when we were plainly still only developing boys by comparison. The main criteria used to select them for promotion seemingly had been prowess at sports and natural leadership, rather than pure academic achievement at either schools or workshops, or plain creeping, which certainly happened in later entries, including my own. They enjoyed a certain reputation, even among the other entries senior to them, that you simply did not mess with them. Shortly before we had arrived, the combined Forty-Seventh, Forty-Eighth and Forty-Ninth entries had attempted to raid the Fiftieth and the Polish apprentices too for good measure, and cause general hell in their barrack rooms, a not infrequent event in the apprentices' world, I might add. Despite the superiority in numbers, they had been remarkably unsuccessful. The resistance had been so strong as to be reported in the national press as a riot. Considerable damage had been caused to both buildings and persons, and at one point the Polish apprentices, known as 'The Rods', with all the determination for which their country is rightly famous, and as the worthy sons of men who were themselves exiles, had taken their rifles from the racks and fixed their bayonets in order to repel the invaders. Fire extinguishers had been dropped on heads from three floors above, and whole window frames were missing, never mind about broken glass. One muscular Fiftieth member, 'Dick', had jumped down the three floors high stairwell on a table top, and on to the heads and backs of the intruders trying to force their way inside. He only suffered a broken leg in doing so, which was generally considered as being a fair price under the circumstances. It most obviously was a rather hard school that we were joining.

There was a certain structure and order within the apprentices' society, of which we had to be made aware and the Fiftieth soon took care of our education in this respect. On one early occasion, 'Slash' Gwilliam was enjoying himself with a newly-found toy. It was the room's mirror, and as it was a bright sunny day he was happily flashing its beam around on other buildings. He had been standing in the end window and focusing his attentions on the window of another barrack block across the way for some time, when the door of our room opened. In came two members of the Fiftieth, both quite large. They walked silently down the room followed by the eyes of all of us except 'Slash', who was still concentrating his attentions on the window of the block from which these two had obviously only recently emerged. When they reached him, one of them reached over and took the mirror away from a surprised 'Slash', tucked it under his arm and they both departed in complete silence. The point had been made and we all soon had to pay for the replacement mirror.

Before we ourselves had mastered the drilling art to a level that would satisfy our mentors, we would see the Fiftieth being marched to work every morning and afternoon from our barrack room windows. They marched in a kind of brooding sullen silence at their own distinct pace, much slower than normal, and were apparently quite deaf to the demands of the drill instructors and the like who snapped at their heels like angry sheepdogs, endeavouring to change the speed to a more service-like one. These minders would regularly move the leading ranks to the rear in an attempt to increase the pedestrian pace but they never seemed able to find anyone willing to go any faster. It was a chilling lesson in how to resist, although, at that time, we still didn't know what it was that we needed to resist.

THE COLOURED HAT BANDS VERSUS THE BLACK HATS

The preacher in a dockyard church
Got up one day and said...

The Fiftieth had apparently decided to formally announce their arrival as senior entry in number three wing, and one day our joint morning parade on Maitland Square was suddenly cancelled when it was noticed that a new flag was flying on the flagpole at the saluting base. It was an article of ladies' underwear seemingly borrowed from the clothes line of the N.A.A.F.I. girls and it had been most securely fixed in place, seemingly quite beyond the strength of the red-faced duty sergeant, still tugging desperately on the rope as we marched past, and who was later requiring the attentions of the station fire engine complete with ladder and crew. It was forty years later that I discovered, in light conversation at a reunion, that I was quite mistaken in my belief that it was the Fiftieth who were responsible and they were in fact quite blameless in this particular incident. 'Ig' Noble from our very own electricians was in fact the guilty party. Not being entirely sure about the source of the garment, our good friend 'Steve' had been given the pleasant task of returning it to those who it was assumed must be its rightful owners when it had finally been recovered, to the accompaniment of a certain amount of rather obvious grinning vulgar enjoyment by those of us fortunate enough to be in rooms with overlooking windows. We soon prudently retired from our observation platforms, however, when the withering upward look and clenching fingers indicated all too clearly that this was most certainly not a subject for humour of any kind.

The Fiftieth ran a very democratic society of their own, quite separate from the one established and maintained by the R.A.F. In times of crisis or whenever they felt the need for some light relief, they would announce a meeting in the N.A.A.F.I. canteen and we, as junior entry, would be precipitately evicted. Leading this society was Reg Cresswell, nicknamed 'Ianto', who had two deep sinister scars running down both his cheeks. There was much speculation amongst us concerning just how he might have got them. He also had a younger brother Ken, who was a 'snag' in the same entry. Ianto had not been promoted to an N.C.O. apprentice, but if ever there was a natural born leader then it was he, and this, mark you, was within an entry which had an abundance of leaders. Just occasionally an entry would produce one of this kind, Jackie Parr from the Fifty-Third entry was just such another example. The service never recognised 'Ianto's' very real talent, nor forgave him either for his irregular influence and, on my last posting before leaving the service, I came across him again, still a lowly corporal but as rebellious and colourful as ever. After the meeting, the pipe band would play, and usually we would be visited and our beds tipped over occupied or not, just as a general reminder of who was who in the apprentices' hierarchy.

Our entry continued this tradition after the Fiftieth had graduated, and we often had sing-ins reluctantly led by general consent by Tom Atkinson, one of our airframe fitters and soon, the lead trumpeter in the apprentices' pipe band. The junior entries would be evicted just as we had been evicted in our more humble days, the serving bars closed, the doors locked to prevent any spying N.C.O. from gaining access and the meeting held. The young girls who served in the canteen would stay out of sight even if they could not get out of earshot, since many of the songs were hardly of the 'Nymphs and Shepherds Come Away' variety, and therefore not exactly suitable for young ladies' ears. Those readers who are of a more vulgar turn of mind should have little difficulty in expanding the appropriate phrases located at the beginning of most chapters. However not quite all of the songs were obscene, and my guess is that they couldn't help but enjoy Tom leading 'Oh You'll Never Go To Heaven', for example, and some of the words of which I have used in the dedication. Some of the natural extroverts such as 'Ernie' Baldwin, a tall Mancunian, took little persuasion to get up and render such general favourites as 'The Hole in the Elephant's Bottom,' 'There was an Old Monk' or 'The Keyhole in the Door', with the masses joining in the chorus, but in order to encourage more reluctant soloists from the audience, a group chant from one side of the room would start up once an individual had been nominated by one of his 'friends'.

"Sing! Sing! or show us your ring", they would chorus, indicating that the poor nominee could either oblige them with a suitable melodic rendering or remove his trousers and show them the round, brown and hairy part of his anatomy. A responding chant of "We don't want to see your ring, so sing you bastard, sing," would erupt from the other side of the room to further encourage the unfortunate, and occasionally still lemonade from half-consumed glasses would be directed from the audience at the naked rear parts of anyone who so chose. Most settled for the song option, I might add.

At a rather later time when we had matured somewhat, and a certain degree of healthy group cynicism had set in, this rather got out of hand one general service training – better known as G.S.T. – afternoon when we had been assembled in the 'tank' to hear one of the periodic addresses by our squadron leader, ex-air gunner padre. This was an unusual and most welcome respite from our normal level of activity on such afternoons, and was gratefully considered by all of us to be a real 'scrounge'. He stood on a small platform at the front waiting patiently for quiet with head bowed, and hands clasped loosely together in front of him, in the general at ease position of his calling. Then someone, probably anticipating a likely spot of early hymn singing, decided that even on this occasion, the normal protocol of the location should still perhaps prevail, and the chanting began from one side of the canteen with growing volume, to be taken up with the usual response from the other. At first he simply could not believe his ears, and he just stood there with his mouth open, staring at the direction from which the opening chant was coming.

"Sing! Sing! or show us..." it began. When the response of "We don't want etc." started from the other side, his head turned to follow it with the same blank incredulous stare prevailing, but not for long. Perhaps it was when they came to the part which raised the question of his legitimacy that particularly irked him. What I can confirm is that he obviously neither wished to give us a rendering of his own particularly favourite hymn, nor to remove his trousers and reveal his rear anatomy to us, because he began to wave his arms frantically in the air and shout, "Out! Out!" and he drove every last one of us out from the 'tank' like a flock of geese ahead of him, and left us stood at attention in the road outside while he went to seek help or divine intervention. The rest of the period was devoted to some extra punishing drill, supervised by none other than the regiment sergeant, who always seemed to hold himself available at a moment's notice for such enjoyable extra curricular tasks.

The normal daily English used to communicate between each other had deteriorated to a level which would have turned a docker's ears red, and the transition to it came about in days rather than weeks once we had been inducted. It was in fact some months before most of us had any use at all for five-letter or greater words, and it required quite a good ear to be able to pick out the odd noun or so in any normal conversation. This was accompanied by a simultaneous degree of vulgarity and coarseness beyond belief in ones so young. Being grown up, your own man, and away from home for the first time had some small privileges after all, it seemed, and certainly one of these was the privilege to use bad language in the absolute extreme.

There was additionally a whole new extra language to be learned. A leading apprentice was a 'snag', an R.A.F. policeman a 'snoop', the N.A.A.F.I. canteen was the 'tank', a sandwich was a 'wad', a forage cap was a 'foo', and a peaked ceremonial hat was a 'bull'. Thus, making unauthorised changes to one's ceremonial hat to improve its appearance became 'bashing your bull' and was very likely to get you punishment, as we all soon found out after attempting to improve our appearance by paring away the hard back shaping with a razor blade, and removing the peak stitching. A legitimate but easy duty was known as a 'skive', and one who found himself on such a duty was enviously called a 'skiver'. Anyone with an inclination to find ways of avoiding unpleasant duties was known as a 'skate' and light easy duties were a 'scrounge'. Thus the early band devotees were known as 'drums skates'.

The Air Force never was very consistent in its standards and while 'bashing your bull' got you into deep trouble at Halton, it was encouraged at Cranwell where our radio and radar colleagues were being trained. We had one of these worthies transferred back to Halton for retraining in another trade after a year or so, and there was much speculation amongst us concerning what precisely would happen when his ceremonial cap went on parade for the first time with us. It was a piece of pure art, more akin to the soft hat worn by American aircrew in films than any item of airforce equipment that we had ever seen; massively arched like a cathedral entrance at the front and drooping low around the ears at the sides like a Lancaster bomber with its flaps down. In the event, the first one to spot it on parade as we formed up for a passing out rehearsal one day was 'Steve' and, what is more, from a distance of several hundred yards. His fingers flexed and reflexed in double quick time and he bellowed like a bull caught by a delicate item in a barbed wire fence as he rushed over the intervening distance at an entirely unwise pace for a man of his years. The poor creature crowned by this masterpiece in elegance was literally dragged from the ranks and dispatched to spend the morning on the more odious of kitchen fatigues. On balance I think he came out fractionally ahead of the rest of us that day. The art for the rest of us now was to attempt such changes as could barely be perceptible on inspection but could be adjusted to more elegant lines, by the addition of manufactured items called 'props' and the crushing of rather than removal of hard padding, just prior to being in the presence of females. We all soon gave this up as being a rather futile exercise, firstly because our minders were much taken in those days with removing your hat and even dismantling it in their efforts to find something wrong with you, and secondly because we were never remotely close to any female. Some lessons just came home quicker than others in those early days.

BUT BABY IT'S COLD OUTSIDE

Standing on the bridge at midnight
Throwing snowballs at the moon

1947 had a very cold winter indeed, and there was a great deal of snow in the south of England. The Fiftieth in customary fashion, followed by us, decided to enjoy it to the absolute utmost and, after the inevitable large scale snowball fights had begun to pall, something more exciting was sought. Located between number three wing and the R.A.F. Princess Mary's Hospital close by was the former site of a temporary wartime camp, long abandoned and the sole surviving building was an ancient corrugated iron construction, which had once been a N.A.A.F.I. canteen and was therefore known as the 'tin tank'. It was derelict and on the point of falling down, but in the event still had some use, providing building materials for the various designs of high speed sledge which began to appear. Taken up the hillsides of the Chilterns that rose up just behind our barrack blocks, even the most simple piece of corrugated iron with the front bent up made splendid one and two man bobsleigh vehicles for extremely hazardous journeys down through the roads of the camp, assuming, of course, that the steering managed to keep the vehicle on the road. In the case of 'Pash' Page, one of the Fifty-Fifth electricians, the steering was either not up to standard or more likely non-existent, and he was soon in the R.A.F. hospital with a broken leg, being visited by his doting mother and father and surrounded by his cynically amused fellow apprentice room mates. The whole of the corrugated N.A.A.F.I. edifice simply disappeared over the brief period of one weekend. All that remained were the wooden stringers that rose like witnessing fingers to the sky. The authorities suddenly discovered great architectural merit and, what was even more significant, financial value in the building, and we all were required to pay for it, despite the fact that it was never replaced.

The inclement weather continued, and with insufficient fuel available for the whole camp, a rationing system was eventually implemented. The barrack rooms were freezing cold and it was almost impossible to get warm enough to even sleep. 'Horace' Wallace and I had the rather original solution of pulling our beds together and using the combined blankets sideways rather than lengthways which meant that we could both have double the thickness of covers available for one. This we did and were of course in bed fully dressed trying it out, great coat, pullover, cap comforter and all, seeking warmth one evening when 'Slash' Gwilliam's corporal apprentice elder brother John, from the Forty-Seventh entry, came over from number one wing to pay a social call on his younger brother and to see how he was getting along. He was of the tall, athletic, possessor-of-natural-authority type, the centre half in the apprentices' soccer team, and very intimidating. As he was leaving he stopped, stood and looked at 'Horace' and me, both happily ensconced in our small comfort, very long and very hard, and at Terry Thornton in the next bed, similarly clad, and busy as ever with his latest piece of knitting. He was most plainly not impressed with the younger generation to put it mildly, and was obviously extremely fearful for his young brother's future moral welfare in such company. Neither 'Horace' nor I, both confirmed heterosexuals, I might add, had quite taken that particular aspect into consideration in our plan, but we realised the implication well enough now, and by mutual consent the beds went back to the frozen single variety the very same evening.

'Horace' had still not quite abandoned thoughts on how to stay warm, however, and took to wearing his pyjamas permanently as an extra insulating layer of underclothing beneath his uniform. That was until the morning parade when an extra-vigilant inspector noticed the blue and white striped flannel material peeping out from the leg of 'Horace's' trousers. A few extra fatigues sessions served to restore 'Horace's' body temperature after that.

The winter of that year was so hard and heating fuel in such short supply that the situation eventually deteriorated to the point where we were all sent home on emergency leave for three weeks when the sanitation systems in the barrack blocks eventually froze solid. I returned home in the happy thought that surely this was at last prodigal son time, and carrying every last single item of my kit to show my new-found standing in the world to the family. Strangely, no one seemed to have missed me noticeably in the couple of months since my departure, nor were they very impressed with either my possessions or my new self. When I returned my A.T.C uniform to its owners, my old comrades from my pre-service days looked me up and down and even gloomily remarked how badly my new uniform fitted. I must admit that I was somewhat downcast by this unexpected turn of events, and the experience was thoroughly rounded off one evening when I got on a standing room only bus wearing my peaked cap and a drunk proffered me a sixpenny piece and requested a ticket to the Old Market Place. Things cannot get much more basic than that, and I was in fact almost anxious to return. After a couple of weeks of time off, and when the weather had abated a fraction, we finally all returned to resume basic training with the happy thought that now it would just go on for a little longer than had been originally planned.

The only one not to benefit from this spot of extra leave was poor Robin Berry who was in hospital at the time suffering from appendicitis. When recovering from surgery, he found little else to amuse him for those weeks of our absence than the snooker table in the games room, but such was his devotion to improving his skills that he was even banned from its use by the nursing sister who, it seemed, not only feared that he would remain permanently bent forward as a result of his devotions, but that he was perhaps attempting to 'work his ticket' and get himself an early discharge from the R.A.F. The likelihood of this since he had only been in the service for barely a few weeks did not seem to occur to her. Perhaps even Florence Nightingale didn't like snooker overly much.

Yet another new experience awaited us on our first morning of return from leave, an event known as an F.F.I., meaning free from infection medical check. One was required to stand to attention beside one's bed – but then again when was one not? – dressed in P.T. kit whilst the Medical Officer, accompanied by the Regiment Sergeant, examined you for anything which one might have been lucky enough to catch during one's absence. In this search for and rooting out of disease some had to merely reveal the palms of their hands, others their hairy armpits and others to lower the shorts and reveal even more private areas for closer perusal by the medical man and his supporting act.

'Horace' was merely asked to show his tongue, and a modest half-inch was offered between closed lips. More was then requested and the length was adjusted to a more generous full inch at which point the sergeant lost his patience and began to bellow at poor 'Horace', who at this angry behest, finally fully opened a mouth most generously naturally endowed by nature and released a tongue which was more like a roll of red linoleum, which apparently impressed the sergeant if not the medical officer no end, because he then loudly remarked with great pity, "Horeed lad Horeed." For some time after this 'Horace' referred to himself as 'Horeed Horees'.

At this time you can imagine how attractive it was to be called outside in P.T. kit, which consisted of a vest, a pair of shorts, and a pair of gym shoes, into those still arctic conditions. The instructors with ever soft hearts, permitted us to wear our woollen gloves, which somehow failed to completely meet the needs. The more enterprising soon began to seek hiding places in the warmth when the P.T. whistles began to sound. The only possible place was the drying room, but this was a far too obvious location and one of the electricians, Paddy Swoffer, eventually sought refuge one day in the loft. He was one of those who had received an unfortunate nickname and had become known as 'Swing Ring' Swoffer, from his rather small stature and his habit of having his small pack hung on a long strap so that the pack was always banging against his bottom. The last that I heard of him in the service, he was a flight lieutenant flying with the Black Arrows, a forerunner of the famous Red Arrows acrobatic team. The loft was above the ablutions that adjoined our room at the top of the barrack block. It meant that he required some assistance to reach the ceiling door far above him, but he finally managed it before the warm sweater-clad instructors evicted the rest of his shivering friends from the safety and comfort of our barrack room. It seems that, some time after the rest of us had left, a group of service policemen came round looking for just such evaders as poor Paddy. With the naturally suspicious nature inherent in their trade, and with the advantage of greater height, they also went up into the loft. Paddy was then obliged to climb into the water tank which supplied the washing needs of all the six rooms in the block and to almost submerge himself in order to escape detection. When we frozen few returned some time later it was to view a dripping wet and even colder comrade emerging from the loft. It was the cause of great laughter and merriment over the next few days until the moment that Paddy confessed he had been so cold that he had been obliged to answer a call of nature whilst in there. We had been washing with and drinking this since. Suddenly the humour seemed to evaporate from the situation. After the emergency leave, 'Horace' Wallace had returned with both a rope ladder and a sliding bolt which he fitted to the loft in order to supply some security and a greater degree of convenience. Always there with a thoroughly thought out plan was our 'Horace'.

It was about this time that we discovered that the central heating system which heated our rooms was carried round all the buildings of the wing in tunnels which contained the pipes. There was very little space, and in places it was blocked off with wire fencing but for an adventurous boy, dressed only in an overall, it was possible to crawl to almost any building, including the cookhouse and even the sergeants' mess. Stolen toast and jam then came on to the Sunday morning breakfast menu for those wishing to remain in bed for a little longer, after 'George' Hanchett and I pioneered the first route to the cookhouse. There was initially a certain element of risk in the tunnel ventures, in that you could only crawl forwards with any ease, and you never knew where you were going or if it might come to a dead end, but in time the routes became as well known as classic alpine climbs. It was some years before any of us needed to gain access to the sergeants' mess for any purpose other than fatigues, but our perhaps unusual route to the cookhouse helped to sustain our growing appetites when our food intake continued to be inadequate, and was obviously, in our view, being unnecessarily restricted by the authorities in an ongoing attempt to remake us in the lean and mean mode. Our fellow electricians in the overspill room had made their own pioneering journeys by the same means to the still empty and locked barrack blocks remaining in the wing. There they had discovered, among other things, that the metal drawer in the empty wall lockers was of the exact size to fit in our webbing back packs, and made a most precise and suitable former. As a consequence most of these empty rooms had been completely stripped of these most desirable items before too long. When this dreadful and shocking state of affairs eventually came to light when the blocks were reopened, new means were then rather urgently required to avoid detection on the inspection to detect the culprits which soon followed. Whilst those who were unfortunate enough to be located on the lower floors were being discovered red-handed, there was just time for those located on higher floors to improvise a solution. The trick was found to be to have the open surface of the drawer to the front of your pack so that when tapped by enquiring knuckles it did not give off a revealing metallic ring. If it did then you were soon in front of your Commanding Officer and even sooner on jankers.

It is simply amazing how boys of that age in the collective can find so many things on which to waste their inexhaustible energies. Another most interesting pastime was the lighting of farts, with a competition for the longest flame measured by minute pencil marks on the room's scrubbed table. The necessary cry of alarm, frantic rush to the table and dropping of trousers at the first significant rumbling of the bowels had to be co-ordinated with the simultaneous arrival of a qualified marker, complete with poised pencil, and a reliable lighter or box of matches. This was abandoned by the electricians when a match was held with trembling hand too close to Curwin Lewis, and he suffered badly singed hairs on his bottom. He was obliged to treat his injury himself with twisted neck and hand mirror, because no one else was prepared to help him, and there was simply no satisfactory explanation of how it had occurred if he chose to report sick. In any case, from the first day there had never been any serious competitor for 'Slash' Gwilliam, who could produce lively explosions literally on demand. At anytime and anywhere, someone could shout, "Give us a real rasper Slash," and he would lift one leg in the manner of Charlie Chaplin, frown in concentration for a moment, and then oblige.

DON'T SAY SOMETHING, JUST STAND THERE

Oh his name was Henry Hall, Henry Hall
And he'd only got...

As the snow at last subsided and spring arrived, a great deal of time and energy was devoted to water fights . It became hazardous in the extreme, to venture anywhere near the washing and bath areas that adjoined and separated the rooms on each floor without risking a sudden drenching from a well directed fire bucket held by some lurking sniper. One weekend this was going on like some running skirmish and had spread to the hallway before it was realised that the water was flowing under the door of the outer room occupied by some unknown sergeant. A peep was ventured to see if it might be safe to put the damage right before the owner's return. Unfortunately the instrument-maker instructor occupier was not absent but only taking an afternoon nap on his bed, and it was then necessary to attempt a mop up on all fours as silently as possible. The astonished owner awoke before the job could be completed, however, and he was discovered to be without any great sense of humour. The culprits were then found other duties to entertain them for the rest of that weekend.

Any concept of privacy had long since disappeared. Even the slightest deviation from the norm of the mass was immediately noted and interpreted by them. The peculiarities of one's body were by now common knowledge to everyone in the room and well beyond, after having been stripped of all garments by the howling masses, liberally painted with boot polish in all kinds of private places and dragged from the room and thrown into a cold bath on so many occasions. It was hardly possible to have a quiet moment alone in the toilet without a grinning head appearing over the top of the door to see if you might be up to something or other with which to taunt you later, or to observe at firsthand the results of the bucket of water which had just been thrown over the door top. There is no experience quite like that of sitting on the pan with your trousers around your ankles watching the water drip from the end of your chin to induce a degree of deep humility within you, I recall.

Pete Deverall was in those days the kind of chap who still did his best to keep himself rather much to himself, and had chosen to distance himself from the queuing for morning ablutions. He had developed the trait of getting up early, dressing at once and going for his breakfast. There was considerably more space and quiet when he got back to carry out his washing and shaving, after all. Even this innocuous act did not pass unnoticed, and so one morning, as he rose briskly and started to put on his socks as usual, he noticed three others who normally lay in their beds to the very last possible moment proceeding to dress at an even faster pace. Pete recognised a challenge when he saw one and speeded up his dressing in response, but was still last of the group out through the room door. The rest of us all rose and gazed from the vantage point of our windows. There was no actual breaking into a run, of course, which would have been tantamount to cheating, but the pace was extreme just the same. Pete had moved into second place, by the bend in the road to the cookhouse, when he passed from our vision, but apparently was eventually impeded by some late arm throwing and elbowing which actually placed him last at the servery, we later were told. He gave up after that and queued with the rest of us for his morning wash. Yet another piece of individual inclination had gone out of the window forever.

As I lay on my bedsprings one Sunday morning, I was surprised to see a broom twisting end on end as it still went upwards past the window opposite me three floors up. On investigation I discovered that it was the Fiftieth entry drum major, 'Spike' Goodrich, at practice in the area between the barrack blocks. As you can see, this was no small height to throw a makeshift mace but 'Spike' took great pride in this particular manoeuvre, and in the fact that, even in the Brigade of Guards military bands, the drum majors never attempted to carry out anything so flamboyant and risky. In the Lord Mayor of London's annual parade earlier, 'Spike' had endeared himself to the watching crowds and got a mention in the national press by throwing up the mace between the trolley bus overhead cables and with never a slip.

To celebrate the opening of the third apprentices' wing the pipe band beat retreat one spring evening on the wing parade ground as part of their rehearsals for the Royal Tournament. I still have a photograph of the event. The band had a goat as its mascot which was named Lewis, the name being derived from the initial letters from London, England, Wales, Ireland and Scotland. The beast was white and it required a good deal of care by its apprentice handler to keep it in parade ground appearance, since its natural inclinations were to be nothing of the sort. When taking part in ceremonials, it carried more silver about its person than a duke's dining table, and it loathed the smell of white blanco even more than I did, to the extent that it would attack anything or anyone wearing a ceremonial belt painted with the stuff, given the slightest chance. It had a pair of long curved horns, a mean quick temper, was not house-trained and, furthermore, was not a great respecter of persons, so it was as well to preserve some distance if it happened to come near you when forming up for some ceremonial parade, or you could end up buying a new pair of trousers. On the other hand the animal was inclined to a natural laziness, and extremely reluctant to move whenever the band got on the march. Vic Parsons from our entry, who was the goat handler for a time, said that personally he used to wrap a piece of lead strip round his fist and punch it in the ear to get it jump-started.

The coincidence of the goat's name being the same as our electrician friend with the singed bottom did not escape our notice and poor Curwyn became known as 'Goat' for a while. He only lost this name when it was discovered, in one of the early cold water immersions, that he only had one testicle. From then on of course he inevitably became and remained known as 'One Ball' Lewis, even forty-plus years on. I realise that, fortunately for me, this must have been noticed prior to my own first ducking, since I was lucky enough to remain plain old 'Grim'. Since poor Curwyn was also the possessor of continually troublesome sinuses and a most ferocious stutter, one wondered even at the time how on earth he had ever managed to make his way through the induction physical medical examination at all. Much later I realised that precisely the same thing could easily have been said about my own psychological tests. All in all, the reliability of those seemingly highly demanding induction parameters became less and less comforting over the years. To make matters even worse, Curwyn was additionally under the grave suspicion of his room mates at that time, since he was being regularly invited to tea at the married quarters of one of our physical training instructors Flight Sergeant Wingate, known as 'The Muscle Man', who was somewhat less than appreciated by most of us. On reflection, what with the singed bottom, perpetually blocked-up nose, speech impediment and partial lack of vital equipment, 'One Ball' most obviously thoroughly deserved his tea.

The sole personal luxury item enjoyed by the electricians of block 7 room 5 was a portable gramophone owned by one of our number 'Curly' Coppack, which he brought back from his emergency leave to help lighten and brighten our young lives. 'Curly' had apparently enjoyed the close companionship of university students before joining the service, and this had, at least in his own view, made him something more of an enlightened man of the world than the rest of us. Despite his apparent interest in something as sophisticated as music, unhappily he only possessed two records, a melancholic orchestral piece, 'The Last Spring' and 'The Boogie Woogie Bugle Boy from Company B' by the Andrews Sisters which were repetitively played according to general mood. More and more as those basic training days passed, there became less and less to be happy about, and the nostalgic notes of the former became a more natural choice than the howling of the latter. We had already had more than our fill of bugle boys by the time that we had returned to our room after another long arduous day. By the time that our first summer leave finally arrived, we all begged him to take his infernal machine home and under no circumstances to ever bring it back again. In our final year it did reappear, but now in a 'rook' barrack block since the owner had by then been promoted to a position of some authority. The records in the meantime had apparently changed to Nelly Lutcher's 'Fine Brown Frame' and 'Come on a My House.' So much for the benefits of the close company of university students.

Our education was being updated in so many ways. The word cleanliness took on a whole new depth of meaning previously not even suspected. It, for example, meant that the packs and straps of a host of webbing equipment, originally Air Force blue in colour, had to be repetitively scrubbed and bleached until they were white, all in order that it could then be blancoed back to its original colour. You could literally spend hours on the brass fittings alone. The amount of attention required to maintain the simple barrack room furniture in its required condition made the original purpose of the items totally irrelevant, and it became vitally necessary to obtain everyday use copies. It is one thing to keep washbasins clean, and yet another to make it necessary to unscrew the metal around the stopper hole so that a cloth on a stick can be inserted to clean at least as far down the waste pipe as an officer's white gloved finger might attempt to reach. People had to learn to protect their room job in order to avoid punishment, and you risked actual bodily harm if you so much as washed your hands after eight o'clock in the evening with an inspection due the following day. The dusting of the outside landing had to be co-ordinated to proceed floor by floor starting at the highest, to avoid the fire buckets' water on a lower floor being contaminated by the dust from the floor above. Such specks of dust floating harmlessly in the water would be described as filth on inspection, and result in punishment. To sit at the wooden table to write a letter would most assuredly provoke at least a comment from the one responsible for its cleanliness that you had better take good care not to get a single ink spot on the surface. It required willing co-operation to carry the one responsible for lampshade cleaning on your shoulders and spare him the repetitive climbing. The beneficial side effect of this was that one soon learned to become less selfish and we began to realise that, in order to survive, we would obviously need each other.

Inspections became battles of wits which we regularly lost in those early days. They could think of new dodges faster than we could keep up with the old ones. It took some time to learn to think like they did, if think is the right word and to anticipate the most outrageously stupid piece of pure bullshit that such a mind could possibly devise. Their all too obvious hope was that we would learn to discipline each other, but we soon passed beyond that sort of thing. After all, this week it might be your unfortunate turn to be caught out in some way and a general punishment awarded, but next week it would most certainly be someone else's so there was not much point in getting angry with the individual. Once we had adjusted our philosophy to take joint action, then things improved somewhat and the punishment rate soon began to drop. The dust in the coils of the bedsprings and the polishing of the brass base of any unscrewed light bulb were anticipated rather than distressingly discovered in this way, and 'they' actually began to take pride in the fact that we were beginning to shape up. It became all too obviously necessary, for example, to regularly unscrew the metal threaded rod which passes through the centre of one's water bottle cork in order to get behind the metal cap and into the threads of the securing nut, and ensure that they did not show even the faintest taint of rust from any water. The prongs of one's fork were soon acknowledged to be potential fertile breeding grounds of disease, but could not be cleaned with metal polish however. The cookhouse tank of water through which steam was supposedly passed for the cleaning of one's eating equipment was quite useless of course after the first few minutes of use, when it had invariably become reduced to a bath of cold greasy scum. The sand in the fire buckets was a much safer cleaning material provided, of course, that you raked it over afterwards. Even the inside of the handle of one's china mug was an extremely dangerous place if the glazing happened to be ever so slightly chipped or cracked. On discovery it would be smashed on the floor as a preliminary to actual punishment. As the eyelet holes in one's boots began to expose the brass under the black enamel through with wear, it became vital to polish them too. The linoleum floors on which we had carelessly scuffed our civilian shoes barely weeks ago came in for literally hours of polishing. Relief teams took over as others tired themselves to a sweat-dripping exhaustion by swinging the heavy metal 'bumper' wrapped in 'obtained' blanket pieces. More blanket pieces were stored just inside the door for use under feet, and one moved about the room like an ice skater on these floor pad savers. Even to be caught in possession of these necessities in those days could result in severe punishment, so that those who slept near the door had to store and hide these items in their own kit space somehow.

The 'they' to whom I seem to be making constant reference were the drill instructors, physical training instructors, disciplinary clerks and R.A.F. regiment N.C.O.s who controlled, and I mean controlled, our existence in the wing area. One common deficiency of these cretins, and there were many, was their complete inability to pronounce recognisable English without extending the vowel sounds. Thus Wallace became 'Walees' and Bristow became 'Breestow'. From now on I will use the phonetic pronunciation which they affected when recording their dialogue, in order that the reader does not miss the benefit of their singular intellect. Their ability at written English was not noticeably better, and continually threatened to wear out the tip of their index finger. I became Briston, Wallace became Valance, and 'George' Hanchett became Hansnett when our names were read from any written document. At any one point in time I could be Grim, Breestow or Briston depending on who was saying it, and if it was written or spoken.

ALWAYS TAKE GOOD CARE OF YOUR WEAPON

She's a big fat girl,
Twice the size of me

When our foot drill had apparently reached acceptable standards, we were issued with our own rifles and bayonets to care for, and we started all over again with rifle drill. The bayonets were of the World War One type, long and sword-like, and of course that meant a considerable amount of burnishing was needed in order to bring the blade to the brightness of the mirror which 'Slash' had recently lost for us. The woodwork needed sandpapering to a silk-like smoothness before finally French polishing made it a work of art. The long leather scabbard required every bit the same amount of attention as one's boots. Just as we had succeeded in these fine masterpieces, they were taken from us and we were issued with the newer spike type bayonet, admittedly with a black metal scabbard which required no other attention than avoiding it being scratched, but which needed its blade burnishing all over again. Even with this level of attention to the purely physical, you could be caught out by a sudden and unexpected demand on the mental level to state your rifle number. It was AL 9358 by the way, but if you were not sufficiently quick enough in your reply then you were in trouble. The rifles were kept in locked racks in the barrack room and, to prevent even the daily dust from settling in the barrel, you needed to keep a loving eye on it. Well, at the very least a piece of four by two partly blocking the muzzle, but loose enough to remove quickly at one pull and be hidden in the hand if there happened to be a sudden inspection. A 'dirty' rifle would get you instant punishment, and random checks at all hours were frequent. We had more items of kit than regular airmen too. In addition to all the normal issue of webbing equipment, we had extra belts, bayonet frogs and rifle slings, to be made gleaming white for ceremonial parades.

The original purpose of the rifles and bayonets was not forgotten, and we spent long periods on general service training playing soldiers. We marched with them on our shoulders, ran with them through obstacle courses, crawled with them under barbed wire and stripped and cleaned them incessantly it seemed. Our good sergeant had been inducing us to make really bloodthirsty shouts as we made bayonet charges which culminated in sticking the bayonet into suspended sandbags during one session, and, understandably in those early days, one felt somewhat self-conscious about making such public and primitive displays. In consequence the noise had been somewhat muted and had not pleased our instructor in the least. We were made to repeat the charge this time with him in close attendance to note who precisely was shouting exactly what, and just how loud. We had not run far in my group before we were stopped by the usual loud scream and an angry dialogue was begun with 'Swing' Swoffer.

"Just what is eet that yew are shoutin', Swoffer?" he angrily enquired at the bellow. "Let us orl ev the benifeet of earin' it yet agen."

'Swing' then screwed up his boyish face into what he obviously considered to be a mask of supreme evil. In fact he only managed to look rather constipated. He repeated in his impeccable public school English.

"I am shouting, sergeant, watch out you rotters, here we come. My friends and I are about to treat you very nastily indeed!"

The sergeant grimaced, shook his head and reluctantly gave up.

"Just make gargling noises then, Swoffer," he advised.

In order to sustain such a high lifestyle, we were all paid the princely sum of ten shillings per fortnight in our first year, with the remaining eleven shillings held back to cover haircuts, old age pension, kit replacements, boot repairs, damages and the like. Any remaining of this was paid to you when you went on leave. Twice a year we were also additionally blessed by receiving an extra shilling for the purchase of the Halton magazine. 'Ig' Noble once bravely tried to depart the pay table with his extra shilling intact, but made barely a few feet before he was stopped, his fist opened, the shilling removed and replaced by the magazine. After the N.A.A.F.I. incident there was a large hole in my reserves before I had even got properly started. The weekly amount was almost totally taken up with my needs for boot and metal polish, blue and white blanco, soap and toothpaste. In the very early months, particularly after a more than usually arduous day, I would sometimes comfort myself in 'the tank' with the odd iced slice or cream doughnut, washed down with a one penny still lemonade, but that sort of luxury was now firmly out of my economic reach. One had to learn to make very prudent shopping for any luxury items. Single portions of chips, for example, were priced at one and a halfpence whilst double portions were charged at three pence. It was common and vital knowledge that it was infinitely better to order two singles rather than one double, however, since due to the how few chips can be put on any one plate factor, there was, on clear evidence, thirty per cent more in two singles than in one double.

The question of other spending rarely arose, but, for the benefit of those enjoying a private income, it was forbidden to drink or even go into a pub, forbidden to smoke until the age of eighteen and then only with parents' and commanding officer's permission in writing, and heaven help you if you didn't have the permit in your possession if asked; you were also forbidden to go out with girls, not that I even saw one, apart from the N.A.A.F.I. waitresses for those first months. Even to have a brief comforting moment of feminine contact with one of those young ladies, one invariably had to stand in a queue like an orphan urchin with your small coins clasped safely in your sticky hand, behind some older teeth-flashing Fiftieth gallant whilst he leaned across the counter and chatted her up with barely concealed lewd suggestions at his leisure, and heaven help you if you showed any sign of impatience or indicated that you had as much as heard a word of their private conversation. It didn't help our blossoming masculine pride either when perhaps the prettiest of these girls, known to all as 'Bubbles', was overheard to describe us, the young gentlemen of the junior entry, as "poor little buggers". This was some statement considering that these poor girls were themselves under the eagle eye of harsh supervision of a manageress known to all as 'the Dragon Lady.'

After preparing herself diligently for the very highest position in her chosen calling, this lady's life ambition had been suddenly thwarted, it seemed, when Auschwitz was closed, and she had been obliged to accept what was obviously the second best location. She was not taking it kindly either. Billy Broughton and I were favoured by a personal introduction one evening when he was endeavouring to forward my musical education on the 'tank' piano by teaching me to play 'Chopsticks' in a lively duet. I was coming along quite nicely when suddenly a dark shadow fell over the keyboard and we both looked up just in time to snatch our fingers to safety before the cover was slammed down and the instrument firmly locked. I would not have been overly surprised if the key had been returned to her bosom for very safe keeping. Not a word was spoken but the scowl revealed that she was certainly not a music lover and our recital was abruptly terminated for the next three years. On occasion this person, I hesitate to say woman, would appear behind the counter herself to keep an ever-wary eye on her charges, and on occasion to keep her hand in at customer relations, and it was an interesting experience to hand over your one penny coin for a still lemonade and see it being turned over and over under her hard examination to determine if it might after all be counterfeit before being thrown contemptuously into the till. It was rumoured that she was bestowing her favours on one of the corporal instructors. In my view at that time, he must have endured very long hard service on bad stations to have been reduced to this. For the life of me, I couldn't understand what it was that we had apparently done to suddenly make them all hate us so. Later I learned with experience not to question this in any way, but to just give them something solid with which to justify their opinion. In fact the only other civilian females whom I even saw, apart from when I was on leave, were in the school playground that we passed each day on our march to work when we commenced real training. All seemed considerably younger than ourselves at that early time, but young growing male appetites being what they were, before too many months had passed more than a few of them began to show noticeable signs of blossoming promise.

Although I did not smoke at that time, I did of course find it necessary to take the elementary precaution of applying for permission to do so. Already it seemed to make complete logical sense to have some degree of protection for that far off day when I would begin, although, when I attempted to get the necessary written permission from my father, he simply did not seem able to understand why someone who did not smoke needed permission to do so, even after I had patiently explained it all to him in my letter. The list of things which were forbidden grew daily until it became easier to remember the few things that were allowed. Proceeding anywhere alone was ludicrously out of the question. We formed up and marched everywhere, to meals, back from meals, all two hundred of us. It was getting to the point where it was almost necessary to find at least two others of a like mind, before proceeding to the toilet. To do absolutely anything at all, you required a written authority known as a chit. To obtain a chit required a written request to your commanding officer. This was not a simple 'Dear Sir; Please may I', but had become either a 'Sir, I have the honour to submit for your consideration the following request etc.' or alternatively 'Sir, I have the honour to submit for your information the following report', type of letter, and ending with a 'I have the honour to be, Sir, your obedient servant'. Until you had managed to get the wording precisely correct, the officer didn't even get to see it, and by the time he had, and sufficient time had elapsed for him to reach a decision on the matter, it was frequently no longer relevant. Another of the forbiddens at that time was being allowed out of camp. Personally I couldn't imagine who might still have the strength, but the electricians, by common assent, needed to buy a common room electric iron to press their uniforms, and so had to request permission. It took about two weeks to get it, and then it was for two hours only and for three apprentices, so that they could march down to the village together. From sheer necessity I soon learned the old soldier dodge of pressing my trousers by watering the seams, and sleeping with them under a blanket on top of my mattress. One electric iron between twenty two of us simply didn't go quite far enough.

Even the sick and poor, and as in biblical times there were always those amongst us, were, despite the protection of their chits, in fact excused very little. Those who were the owners of an excused marching chit because of temporary injury or infirmity were naturally of course not required to have to march to work in an en masse entry as the fit and healthy of us were required to do, but if they entertained any hopes of dragging their faulty limbs along in painful solitude then they were soon to be disappointed. They would invariably get no further than Wing Headquarters, flagpole line before 'Steve' would be out of his office to intercept them, form them up into mini squads, appoint someone to be in charge, and the blind, lame, and sick would then have to limp and stagger their heavily bandaged, plaster-casted, crutch-supported or heavily-splinted way down the hill in unison in organised ranks, in pursuit of the same distant pied piper ahead of them as the rest of us.

One of our entry, who was the possessor of many chits to excuse him all sorts of things because of bad feet, and therefore was aptly nicknamed 'Boots', finally submitted a request to excuse him carrying chits. This was rightly regarded as an attempt to be frivolous and he was punished.

Everyone was being punished for something, it seemed. It might be a few circuits of the parade ground at the double, with your rifle held above your head, or an evening cleaning out the cookhouse's greasy saucepans, or perhaps waxing and polishing the sergeants' mess floor. Failure to meet standards on inspections meant doing it all over again and for as long as it took; sometimes it went on in repetitive cycles for whole weekends. Even random casual checks of rooms which revealed some minute deficiency in a kit lay-out would result in the whole of someone's kit being thrown out of the open window. My room was three floors up, and by the time you had walked down and collected it and returned, and then had to wash it all again and polish it and lay it out once more, you had suffered some inconvenience, to put it mildly.

Rationing was still in force in post-war Britain, and children were sometimes entitled to extra scarce luxury items. Being still children in the eyes of the state, if not those of the Air Force, we were also entitled to this privilege, and so one spring day we found ourselves on a parade - where else - to receive our banana chits. After all the forming up and marching back and forth, plus the bare economic fact that I could not possibly afford to purchase the benefit, it seemed a rather doubtful privilege and I was not long in selling my chit to one of those with private funds. 'Horace', however, ever one with an eye for the warming of his soul, and an early desire to fight back, retained his for the sheer joy of noisily devouring all of his quota before the hungry eyes of the ever watchful 'they.' I have seen no gorilla, orangutang or chimpanzee at any zoo since, which even approached 'Horace's' lip smacking performance that day.

One of the really important lessons not taught in any lecture, but nevertheless to be thoroughly assimilated, was that there were in fact two Air Forces not one, and commissioned officers were not members of the same one that we were. The purpose of the 'they', or at least until in the fullness of time we grew up to become a new 'they', was to act as a buffer to prevent any small contact between us that might bruise the sensibilities of our noble commissioned, usually ex-aircrew leaders. Fair is fair - some of us were, after all, plainly working class. In the meantime it was apparent that we were being prepared for an intended life of being there purely for the heavy work, whilst they apparently were there for the thinking bit. This was a most valuable lesson, and an accurate appraisal of what would be encountered in the coming years. In all my service time of fourteen years, I only ever encountered three officers whom you could really totally trust. Their world and Air Force were as remote from mine as Mars or Jupiter. Officers incidentally also have their own peculiar idioms of speech so that you don't have to even actually see them to recognise who they are. It was quite unlike our own vulgar language, with its heavy overtones of sexuality. For them, anything beginning with a letter R was capable of a change of pronunciation. Our 'route' was an officer's 'rowt' and our 'Ralph' was their 'Rafe', for example. One could not help but overhear interesting examples of this lack of a common language and understanding going on around you.

On parade one Saturday morning, we were being inspected by our decorated Flight Lieutenant adjutant followed in close attendance by good friend 'Jim'.

"Sergeant, just look at that apprentice's lizard," he began.

"Lizard Sah. What lizard?" answered a surprised 'Jim'.

"That lizard thing there," replied our noble leader, pointing at someone's bayonet frog.

"That ain't no lizard Sah that ees a frog," explained 'Jim' quietly.

"Well I knew that it was a reptile of some sort or another," came the huffy answer.

Our old friend the regiment sergeant never spared himself in the improvement of our young characters. R.A.F. regiment persons are known throughout the service as 'Rock Apes' because of their service in Gibraltar. I could think of a better reason without even trying in those days. His great delight was drill, any kind of drill at all including foot, rifle or battle, and under his guidance we came to know them all so well. It was all just one big happy business to him and he had an almost magical way of his very own to encourage you in all drill matters.

"Yew are marcheen like a ruptured duck, apprentees," he would frequently advise some poor individual, and his absolutely favourite expression for us in the mass, delivered in a rising crescendo, was, "Montgomery would weep eefe could see you."

One day his drill instruction was interrupted by the passage of a staff car flying a pennant to indicate that the passenger was an officer of high rank. 'Jim' of course called us all to attention and smartly saluted. After the car had passed, he asked us who the occupant had been. I believe that it was Ernie Baldwin who replied that unfortunately he did not know, but his word could be taken that it most certainly had not been Montgomery since he wasn't weeping. Not all of us were as bold as Ernie, however, and it wasn't long before our mentor had selected one of our entry who was prepared to march out in front swinging his arms like some demented windmill and crashing his new freshly hob-nailed boots into the ground at the loud behest of our friend. Eventually he even made corporal apprentice and it was largely for doing this boot-banging business.

There were also some other disturbing signs that some of my fellow apprentices either actually liked this nonsense, or saw in it an opportunity to begin to pursue their young careers. The personalities were beginning to emerge. Some suddenly became religious, and spent much time in the company of the padre at his club for the spiritual and those seeking inner peace. The padre had quite a say in wing matters, we all very soon realised. With my particular recent home background, I was quick to appreciate, with my daily growing acumen, that had I chosen to confide at least a tenth of what I really felt, I could probably have milked the situation all the way up to sergeant apprentice. It was a time of discovering one's own essential character, and having to act accordingly. One of the spiritual in a senior entry had to be moved off camp for his own safety when the Fiftieth passed out. Almost invariably this type turned out to be amongst the biggest bastards that emerged as I recall, so perhaps my moral stance, or was it cowardice, was quite justified.

Others took up model aircraft construction at a club with an officer enthusiast chairman. By the way, a home town friend of mine in a junior entry Ken Barker, later lost an eye in a model flying accident and was medically discharged as a consequence. There was even a model railway society, believe it or not, for those finding it hard to be separated from their 'choo choos'. All the major English eccentricities were fully catered for, including beagling, although coming from a working class background myself, I didn't know quite how one 'beagled'. 'Swing Ring,' already with a deep interest in a future flying career, moved his bottom down to the gliding club in his free time, and began his first year task of manhandling the gliders for the benefit of his seniors in exchange for the promise of some later flying instruction.

Our minds and free time were being diverted into what could be called healthy channels and away from anything which might even remotely rear its ugly head, but, as with all boys of that age, that was of course the only thing in which we were really interested. Some club memberships were taken up with, shall we say, not entirely honourable purposes. One of our entry's more enterprising electricians, with a mature and informed eye to the future, joined the radio club, with the sole objective of having a legitimate purpose for being in the Commanding Officer's office in order to switch on the barrack block radio situated there, realising that in that office lay information, and in information lay power.

Some, 'Swing' included, had arrived with their very own, now forbidden, hobbies such as motorcycling. By all kinds of devious means, they found local places where they could secretly garage their forbidden machines. Two of our entry even went so far as to purchase a motorcycle for the sum of ten shillings from a man who, it transpired, was the local policeman in an adjoining village. Without the benefit of driving licence, tax, or insurance because of their lowly income, and probably on stolen petrol too if the truth be known, they carried out a complete rebuild, then reconnoitred the local county on this machine, as their hobby. They prudently always waved in a friendly fashion if they chanced to pass the original owner. Where gentle personal persuasion failed to indicate the correct route to a career, outside influence could and did succeed. After an unannounced visit to our wing and a private interview with our C.O. by an officer parent, one of my entry became an animal lover overnight and from then on regularly took the C.O.'s alsatian dog on walks without requiring two others to march with him or a special chit either. I openly speculated at that time, that it was perhaps the dog that had to carry the chit rather than the apprentice, and if so where on its person it was located. There is no prize for guessing where I believed that to be. The majority of us regarded this kind of behaviour with some disdain and disgust and stayed well away from these people.

As well as these types, genuine personalities began to emerge and these characters began to come into their very own. One was certainly our already mentioned friend Ernie Baldwin. One summer's day we had been allowed to parade in shirt-sleeved order, and Ernie had decided that he preferred to wear a civilian, highly decorative and absolutely forbidden metal belt rather than the permitted Air Force cloth belt. As he well knew, it was not long before this unusual item of dress was spotted by the disciplinary Flight Sergeant in charge. So it began.

"Baldween Laad. Take off that belt at once. You know it ain't not allowed."

Ernie's reply was respectfulness itself, as he came to attention, "I can't, Flight Sergeant, because my trousers will fall down".

This brought the angry tone into immediate effect, "Don't try and bugger me about Baldween. Do as I tell you this very moment."

Ernie didn't give up, despite the giggles already starting in the ranks. "If they do fall down, Flight Sergeant, then all my mates will laugh at me".

The Flight Sergeant then lost his temper, rushed up to Ernie, all red-faced, and began to try and remove the offensive belt himself. It became pure farce as Baldwin now pretended to be ticklish, and doubled up in convulsions struggling and giggling, "Oh no, Flight Sergeant, stop, oh please stop, it tickles." The noise of this and the roars of laughter going up from the rest of us resulted in the C.O. opening his window to see what the devil was going on. He called the Flight Sergeant inside for a general soothing and calming down. Of course Ernie was punished, but sometimes it actually began to be worth it, since sticking rigidly to the straight and narrow didn't seem to be any guarantee of safety.

This particular Flight Sergeant, known to us as 'Harold', remembered Ernie rather well and singled him out for all sorts of petty discrimination from then on. He had only two tones used in on duty communication. Either he yelped like a terrier which had been trodden on, or when pleased made equally unintelligible low cooing sounds. I rarely heard the latter. He sent Ernie one day for a haircut, under escort by two corporals to make quite sure that he complied. The barber was a civilian who was probably a sheep shearer in some former existence, and a very nasty piece of work indeed. It seems that, after getting Ernie in the chair and making the first few passes with the clippers, Ernie suddenly took alarm, grabbed the clippers and examined them, shouted that they were rusty, punched the barber, and left with his half haircut. The barber could not be persuaded subsequently to even let Baldwin into his shop, and certainly not to complete his half-finished task under any circumstances. So it came about that Ernie appeared on punishment for some weeks with his half haircut and seemed to be thoroughly enjoying the whole affair. Ernie was given a medical discharge after a year or so, to the vast relief of all the wing personnel because he was obviously well on the road to becoming another 'Ianto'.

The civilian employees on the camp were often far worse than the service people. The camp tailor was a typical example. When I had been kitted out, it was he who had decided that I would grow into my best blue uniform jacket, and that he would not need to carry out any alterations. I had not yet managed to conform to his vision of the emerging Bristow, however, and it still hung on me in the fashion of being well fitting on a boy of twice my size. It had not escaped the notice of the 'they' who insisted that I re-equip myself with something more snug. At my own cost, of course, it goes without saying, since the tailor refused to acknowledge any mistake on his part. I was not the first to have encountered the negative side of this individual either. His motorcycle went missing one day and could not be relocated in spite of the combined efforts of civilian and service policemen. I could have told them, as could any other apprentice, that it was now in small parts and residing at the bottom of the local reservoir.

I can say with some personal conviction that the ugliest apprentice in my entry was Ernie 'Jumbo' Parsons, who hailed from Battersea. Sadly he was later killed in a flying accident after eventually turning aircrew. Actually he had missed a very promising career in horror films by joining the apprentices, and would have had the advantage of needing no make-up at all. Of course that alone made sure that he was an early natural victim of the 'they' and he accordingly suffered much abuse with great fortitude. Beneath his fearsome visage, he was a gentle kindly human being, and, perhaps not surprisingly when you got to know him better, he was already an exceptionally talented ballroom dancer. Later on, when the rest of us grew up a little and began to take an active interest in girls and dancing, we found Ernie well ahead in that game, and, what is more, very popular indeed with the young ladies. Actually this gave me some badly needed encouragement because I secretly thought that I was probably the second ugliest of the two hundred. Looking at my early photographs I can still detect a quite remarkable similarity to a fledgling owl.

Amongst the electricians, Cyril Wallace, better known as 'Horace' or later, for reasons which I will shortly make obvious, 'Goolie'

Wallace, was making an early bid for notoriety. He was one of nature's natural extroverts. One of his many favourite ploys, when the opportunity presented itself, was to creep up on the C.O.'s car which was always parked outside his office with the aforementioned alsatian dog inside. Horace would then suddenly stand up straight with his face pressed against the windows and his fingers pulling the sides of his mouth open into a grotesque mask. He would then flick his tongue in and out in rapid motion with accompanying throaty sounds. The dog was of a rather nervous disposition even before all this started, and it simply went berserk every time. 'Horace' always took good care to be well out of sight, before the office window opened to allow the C.O. to see what was disturbing the dog. In time, with patient training, conditioned reflex took over, and the dog would invariably go into hysterics whenever it spotted 'Horace', sometimes at considerable distances.

The essential basic groundwork had been carried out somewhat earlier when 'Horace' had decided on his own initiative to put a bit of a spoke in the wheel of our friend the dog walker, and had feigned an interest in the animal, and, what is more, had given some indication that they might even share responsibilities for its welfare. The walker, badly in need of a friend, was suffering some unpopularity at the time, and was therefore understandably willing to unload at least some of it on to someone else's shoulders. Foolishly he agreed, and 'Horace' made a beginning by setting out with the expressed intention of bathing the poor creature one evening. Fortunately the animal was discovered in the nick of time, in the bath, with the water running and the duckboards tied on top. The exerciser, realising just how close he had come to having to explain the manner of the dog's unfortunate demise to its C.O. owner, was never so foolish as to allow his charge to fall into anyone's hands again.

'Horace' was simply not the macho type. He loathed any form of physical activity, despised all games, and had no love for drill, P.T., or playing soldiers whatsoever. Most of these events were invariably a part of our Saturday morning routine and 'Horace', being a somewhat methodical man, regularly began to take himself off and report sick on Saturdays. He would march to the sick quarters and after the inevitable wait would describe his symptoms to the Medical Officer. As an aid to memory and to avoid confusing the M.O., he had prepared a written list which had such entries as 'April 3rd:

Can't shit.' April 10th: 'Can't stop shitting,' and for periods well in advance. 'Horace' rejoined the routine of the rest of us when his notes chanced to fall from his pocket one Saturday morning, and were picked up and read by an incredulous M.O.

On an early room inspection our C.O. was escorting the Wing Commander who was doing his best to indulge in a spot of putting himself about a bit, and other general good chapsmanship by exchanging pleasant dialogue with the, as ever, rigidly at attention apprentices. He happened to select 'Horace' as one of the fortunate ones to receive his largesse that day.

"Tell me, apprentice, what exactly are your sporting interests?" he enquired.

"Sailing sir," replied 'Horace' which caused some lifted eyebrows among the rest of us since to the best of our knowledge he had never even set foot in any floating transport let alone take an active interest in nautical matters.

"Long way from the sea here, I'm afraid, old chap. Have your own boat and so forth or what?" went on the Wing Co.

"On loan, sir, just a small six footer. A bit cramped perhaps, but good enough for short events," 'Horace' went on. Then the penny began to drop for the rest of us, if not for our noble leader. 'Horace' was referring to his six foot bed in which he indeed did take a great deal of interest, given the least opportunity, that is.

"Good show," murmured the great man as he went on his way absolutely none the wiser.

PUTTING THINGS RIGHT

And we've nothing to eat
'Cos we've thrown all our rations away

The cookhouse was staffed, since the departure of the chocolate biscuit man, with the more ordinary type of cook found on all R.A.F. camps. The rating had therefore dropped somewhat in the ranking list of the latest Egon Ronay gastronomic guide to good eating places. Since we were often employed in carrying out the more menial of what would normally have been their tasks, these cookery persons tended to become a little 'uppity' from time to time, and to not only look down on us but to attempt to impose their will on us too. Why not? After all, everyone else was. We thoroughly detested these 'boggies', short for bog men, short for toilet cleaners. The food was not exactly what it had once been as I have said, and one morning at breakfast, after we had begun training, it chanced to hit a new all-time low. A Fiftieth entry member, exercising his usual right to go to the head of the queue, looked at the slop with disgust and requested something which was at least edible, but this only produced abuse and put-down invective from the greasy aproned creature who happened to be dishing it out that morning. He rather swiftly encountered a punch in the mouth, and found himself on his backside with the plates of food delivered at his head. A sit-down strike then began which ended some few hours later with a stirring speech from the Wing Commander, a promise to improve matters, a return to work still hungry, and punishment for the ringleaders in both the Fiftieth and our own entry, chosen at random from both entries since no one was sure who really was to blame. It was in this way that I learned of the more subtle distinctions between what passes for law and what passes for justice in the Royal Air Force.

   The standard was continually up and down over the years, and in the early hours of one Sunday morning, quite some time after this, a mixed group of Fiftieth and Fifty-Fifth entry came down with a malady which required our instant attendance in the toilets, only to beat with frenzied anxious hands on the doors which were already occupied. It was obviously food poisoning. We were obliged to report sick and to be marched en masse to the sick quarters in order to see the duty medical officer. As we stood outside, nervously hoping that another attack wouldn't come just at that precise moment, he finally arrived, looked at all of us in the ranks, made an instant diagnosis, announced to the N.C.O. in charge, "They look all right to me," and went inside never to reappear. One corporal apprentice was admitted as the test case for all of us, but he was ejected after being told that since only half the wing had complained it couldn't possibly be food poisoning. That was the sole treatment which we were to receive, and pressing needs, shall we say, forced our early return to the barrack block.

The R.A.F. was getting around to correcting those small defects detected at the induction medical. My own experiences with the Medical/Dental branch in those days were not encouraging. The small defects were now being attended to, and I had to visit the dentist. He had as an assistant at that time a W.R.A.F. young lady dental hygienist, who was without any serious doubt the focal point for the animal desires of the whole of number three wing from the highest ranking officer to the most lowly apprentice, such were her obvious charms. It was, therefore, an extremely enthusiastic young Bristow who visited the dentist on that first occasion, even going so far as to anticipate with much pleasure the prolonged treatment that might well be required and just how close certain delectable parts of this young lady's anatomy might be for my closer perusal.

After getting my attention by seizing me by the ears so that my eyes watered, and clamping my head so severely that I could barely see her at all without considerable difficulty, the dentist became most interested in my top two front teeth which had been twisted as a result of a childhood collision with a dog in a competition in which I had taken the silver medal. He seriously proposed that he would remove two teeth on each side of the two front ones, and fit in a brace. They would then be perfectly straight in a few years, it seemed. In the meantime, of course, unfortunately I would just have to look like a chipmunk. When I enquired what about the two gaps on either side so created, itself no simple matter considering the usual amount of stainless steel hardware temporarily residing in my mouth, I was told that he or someone else would fit false ones later. Sadly I realised that with such an appearance I could never possibly hope to win her, and, what was even worse, she might even laugh when I came for future appointments. I was not prepared for that to happen and just left and never went back. The Air Force did a little better with regard to the nosebleeds that I had suffered since childhood and cauterised the inside of my nose, which cured me for all time. I can't say that I exactly enjoyed the moment when the burning taper was thrust up my nostrils but, on balance, a cure is a cure after all, and, whilst I was perhaps unable to smell the roses for a while, I was also freed from some of the less pleasant side effects of 'Slash's' specials.

I had managed to incur the displeasure, quite inadvertently I can assure you, of a certain physical training instructor corporal known to us all as 'Swede', who was soon to be elevated to the lofty rank of sergeant, and as a result he took upon himself a particular interest in my general mobility and physical welfare. This unfortunate situation came about when I became the victim of what I would describe as group humour. All two hundred of us had been enjoying yet another physical training period on the barrack square under his sole control, which had begun as ever with an order to elevate the elbows and take in a deep breath to the full capacity of our lungs. The next order again as always had been to, "Hold It!" I had then been amongst many others, but unfortunately the one at the front in plain view who lowered his hands to cover his penny to the open amusement of the rest of my colleagues, but not to the 'Swede' however. This event greatly enlarged the already rich tapestry of my daily life, I can assure you. Largely due to his unstinted efforts, my condition improved to the point that I could double round the barrack square for a longer time than he could run beside me. It became truly enjoyable to watch him begin to sweat out of the corner of my eye, and still keep on going, just waiting to hear him pant that I could now return to the ranks.

Both entries in three wing were paraded one morning in open order and inspected by two somewhat bruised R.A.F. policemen. No one was identified by them, but one couldn't help but notice that at least three of the Fiftieth entry were sporting somewhat different hairstyles from usual, and there were some quite remarkable changes in their normal postures. It seemed that the previous evening 'Ianto', 'Wee' Sharpe, and other friends had been enjoying a quiet pint in a nearby village pub, when the two R.A.F. policemen had walked in looking for just such wayward apprentices. 'Ianto' had been the first to see them, and had laid them both out at one go, before he and his companions had beaten a hasty retreat.

Eventually our dreaded basic training time was finally over and we began our real training. All of us had anticipated exactly what basic training would be like, but I for one had become increasingly suspicious and aware that the Fiftieth, on normal training, did not seem to be faring any better than we in the wing areas. I began to dimly understand the hidden message contained in their own sullen marching pace much better. They were under the control of the very same N.C.O.s and enduring exactly the same kind of nonsense that we were. 'Harold' was now their squadron N.C.O. with all the joys that this entailed. Thought leads to more thought, and I began to be mildly apprehensive concerning what we might encounter at schools, where we were supposedly to be educated to ordinary national certificate level and at workshops, where we were to be taught our trade. Maybe the Air Force wasn't any better in these areas than it was in the wing and things were not in fact going to get any better at all.

With our release from basic training came certain theoretical benefits, such as being allowed to go to the nearest town, Aylesbury, at the weekends after duty. Without the benefit of money in your pocket that particular joy soon palled. 'Horace', however, was never one to fail to extract at least some small pleasure from the crumbs offered to him, and before too long he began to take an unusual care in his appearance before departing alone for places unknown on sunny weekend afternoons. It was obvious to us cynics that he was, to coin the current phrase, 'putting himself about a bit'. Since he was of the beanpole slender type of Scot, with a set of teeth to challenge the best of tombstones in any decent graveyard, and almost no shoulders on which to support his rifle, none of us held out much hope for him, although we were too kind to put it into actual words. After all who needed more bad news?

In the event he did much better than we gave him credit for, because soon he was invited to tea on Sunday at the home of the parents of this young lady, and on his return we cynics were treated to a slice of his boasting that now he had at last succeeded in getting 'his trotters under the table' as he so eloquently put it. This had apparently been his intention from the very beginning it seemed, and was not to be confused by us with romance or any of that sort of rubbish. This was a matter of pure survival. Those of us who eventually caught a glimpse of her quickly saw his point, and he was treated to some ribald comment in the room after lights out about 'Buck-toothed Betty'. 'Horace' was a more sensitive soul than anyone even suspected at that time, and while he positively enjoyed being the butt of such comments from the mass, he was not inclined to take it from any individual for whom he did not hold at least some slight degree of affection. 'Swing' Swoffer happened to be one of those for whom 'Horace' held no high regard. With his private school-accented English, and signs of obvious breeding, he was not 'Horace's' type of Englishman at all. So it was that one sports afternoon I chanced to be coming into the room as a swollen-faced and bleeding 'Swing' was coming out. 'Swing' had ventured some chance remark about the current state of the relationship with 'good old buck', and he had received a few swift ones in the chops before anyone could intervene. Everyone's education progressed even further forward, in that there are in fact friends, and then there are friends.

We prospective electricians were rather a mixed bunch with all kinds of individual physical builds, inclinations, talents and towns of origin from Perth in the north to Brixham in the south, Maidstone in the east to Ammenford in the west. Pete Deverall, who was gangling and thin to the point of emaciation, was a very promising cross country runner, for example. Good enough to run against Emil Zatopek in the Britannia Shield event, as did engine fitter 'Chas' Tighe the same year that Zatopek ran and won in the London Olympics. 'Slash' Gwilliam was the grandson of a Welsh international and Wolves professional soccer player, and he went on to play for the apprentices' team as had his elder brother before him, and so did Ken Smith. 'Jock' Clark had been a trotting race apprentice jockey and at the other extreme Terry Thornton liked to knit in bed as a hobby, which had already caused some raised eyebrows as I have described. Some, like 'Horace', were inclined to be extrovert in nature whilst others, like Clive Evans and Johnny Hopper, were shy and retiring. 'Dai' Evans was light-hearted, and had unfailing good humour even when in the worst of situations, whilst 'One Ball' Lewis, his fellow Welshman, was inclined to be morose and moody. I was somewhere in the middle with my mouth open a little too much for my own good, I now think.

Summer came at last and we all went home for our first long leave of three weeks. We also had two weeks' leave at Christmas, one at Easter and a seventy-two hour pass at Whitsun and at each mid term between leaves. I grew to hate those seventy-twos over the years at Halton, invariably just short enough to leave you depressed about returning, but on the brighter side long enough anyway to get my socks darned by the females at home. I never did learn to master that particular skill myself, by the way. The longer leaves were different and even gave a certain light-heartedness and joy of expectancy for the weeks before, even if you did have to march en masse down to the railway station at Wendover and wait in entry order for the next train to London. Then you really were able to see just how senior you had become.

Once home you had the time to readjust to being a free and individual human being. It didn't take me long to buy myself some civilian clothes with the little money that I still had in a Post Office children's account, nor to distance myself from the Air Force as far as possible at these times either. The joy of wearing uniform had considerably palled, you see, since the incident with the drunk on the bus. I was noticeably bigger and fitter already, however, to the point that I now almost fitted my new best blue uniform when I could be persuaded to wear it, that is.

There had been a few new developments back home in Cleethorpes too. The second eldest of my three stepsisters had separated from her husband and taken up residence with her baby son at my father's house. The family was growing again, and my father was already plainly deeply involved in it. The R.A.F. in its generosity paid us the ration money which they normally spent on us every day for the period of the leave. It was less than thirty shillings a week. I, of course, offered it to my stepmother since it was rightfully hers, and she took it, in spite of there being three adults in the house, all earning good salaries and knowing full well how much money I had to live on. I made a point after that to always pay my way when at home, and later, when I sometimes took my mates home with me, I paid their ration allowance too from my own pocket. In all my time in the service, I never received a single penny in help, and just one food parcel, and that came from an aunt for my birthday one year, but I did learn a valuable lesson in life: how to keep my pride. Things didn't seem to have changed much in the half year that I had been away and no matter how hard it was, and it certainly was hard, I had been right to leave and find my own way, I decided.

Quite apart from such practical matters, I realised too, when I walked on the beach, and saw and heard the sea again, that for me the Chiltern Hills were something of a poor substitute, and that I sometimes missed those sights, smells and sounds from my very early years. Those roots tend to go very deep, and I have in fact gone on missing them for the rest of my life.

READING, WRITING, AND 'D' CUP

But they gasped with surprise
When they saw the great size...

Until this point in time in our young apprentice lives, it had all seemed very much to us as if we were members of a new junior form in boarding school, and our general behaviour and attitudes to it all had been mirrored accordingly. Every new hardship had seemed to us to be some kind of amusing trial, and our strictly limited world still seemed to present some marvellous opportunities to explore and to exploit. Now things took a distinctly new turn as we settled down to seriously begin our education.

As no longer 'rooks', at least in our own eyes if not in the Fiftieth's, our apprentices' weekday training life began to consist of three half days a week spent on technical education, known as schools, one half day on general service training, known as G.S.T., and which included physical training, drill and military matters perhaps more usual to an infantry soldier than a craftsman. One half day was spent on sports and the rest at workshops learning our trade. There were more than just weekdays for fun and games however. It began every single Friday evening which was devoted to cleaning our barrack rooms to an even higher standard of acceptable brightness than usual, known as a 'bull night'. The echoes of that ringing in my sub-conscious still give me the odd occasional nightmare and even these days just the sight of my wife getting out the basic tools of domestic cleaning can drive me from the house on the flimsiest of excuses urgently looking for a scrounge.

Every Saturday morning was either a wing inspection of the previous evening's labours, or a ceremonial parade or both, plus more G.S.T. with the odd full kit inspection thrown in for good measure as time allowed. Once a month, Sunday was given over to spiritual education in the form of a church parade, which could be absolutely relied upon to occupy the whole morning until lunch time since the church was located even further away from our barrack blocks than the workshops, and one did not simply stroll down there and back in happy family groups, nor did the padre shake your hand or have a quiet friendly word with you individually as you came out either. 'Tosca' Osborne, one of our armourers had declared that he was an agnostic, and on the surface this would seem to have been a remarkably wise decision at an early time, but no one wanted him to be left out of such a significant and meaningful part of our new lives of course, and therefore he was given the spiritually uplifting task of cleaning out the squadron offices during church parades. The rest of the time, however, was absolutely your own, if you still had the strength, that is.

Like our seniors, the Fiftieth, we began to be marched to work every morning en masse, and, on our merry way from number three wing, unlike the entries from numbers one and two wings, who even had the busy civilian traffic held back for them by white-capped service policemen as they marched in magnificent splendour behind the bands past main point, our daily route merely took us past the local junior school. The only ones watching us with any degree of interest at all were what must have been the more advanced members of the eldest girls' class, who were all of twelve years old. One in particular though did have two outstanding qualities as I recall, which belied her very limited age, and the snugness of fitting of her choice of apparel on any particular day became the standard subject of many earnest mid-morning break conversations amongst the many of us suffering badly from a rising but useless hormone production, with much 'ooing' and 'aahing', sharply sucked-in breath and guttural throaty accompanying sound effects. As she grew up on one side of the fence over the next three years so did we on the other, both eyeing each other daily, and I reflect that our increasing entry seniority probably proceeded in direct proportion to her bra size without our even realising it.

Although we could now march more or less in step and were rather proud of our new-found maturity, in those early days our miserable level in the pecking order was even noticeable to these simple children. The more adventurous and bold of the tiny tots, particularly those with a serviceman parent from whom they had obviously learned a thing or two, would come to the side link wired fencing of their playground after the Fiftieth entry had passed, no doubt egged on by their own nine year old seniors. With their little fingers curled through the links of the fence, watching our miserable steady passage of ranks, they would jeer at us in common unison, "Juffs! Juffs!" which was a vulgar abbreviation of those times and place for 'Join Up For F***'s Sake', and normally used to indicate that the subject of attention was a lowly A.T.C. cadet and therefore not even worthy of being a member of the R.A.F. yet. It was most humiliating. Sometimes, just for simple variation, their cry was changed to, "You'll be sorry", and many of my companions including myself, were already beginning to be so on an almost hourly rather than merely daily basis. In the fullness of time, we learned how to make a half-whispered retort from the closest rank to that fence strike home, when the provocation proved too much to bear. Anything louder than a whisper inevitably led to swift punishment being administered by the N.C.O. playing the part of our minder that day, and was therefore hardly worth the effort. On these rare occasions, using the natural advantages of our obviously greater maturity and higher level of sophistication, we would enquire from the corner of our tightly pressed lips, "Have you managed to grow any hair on it yet?" Whether small boy or girl, this usually sufficed to reduce them to a thoughtful silence, which in itself answered the rhetorical question.

As our first summer in captivity slowly faded and finally gave way to autumn and then winter, we began to tread that same old daily path, now carrying hurricane lamps at suitable file intervals to guide our way in the dark like some grossly enlarged glow-worm or group of seven dwarfs, and the daily dull grinding monotony of it all began to grow.

"WHAT DO YOU KNOW, JOE?"

(The correct answer to which is, "Nothing, I'm Air Force Trained!")

Once our real training had started, I began to encounter some very real and previously completely unexpected problems. At schools for example, the boys who had arrived by way of the polytechnic institutes had a clearly distinct advantage over us ex-grammar school boys, who had most obviously and unfortunately been prepared for a somewhat more academic lifestyle. I, as a typical example, had no experience whatsoever of engineering, either electrical or mechanical, other than as a small part of physics. I knew absolutely nothing about engineering drawing, which had a bare nodding acquaintance with my grammar school's subject of art, in that they both used a pencil and in my own particular case more often an eraser, and, apart from mathematics, the remaining subject on the early Halton curriculum happened to be R.A.F. history, which I began to feel that I was perhaps already helping to make in no small way.

My first attempts at engineering drawing soon attracted the keen appreciative eye of our Flight Lieutenant instructor, who stared thoughtfully for some moments over my shoulder at one of my early masterpieces, and then asked why did I not put a few trees and a windmill or two in the background. After that initial mental setback I applied myself sufficiently to learn to at least manipulate the dividers well enough to be able to take direct measurements from the drawing of my more experienced friend 'Horace' Wallace who sat next to me.

In fairness the Air Force was also going through a difficult phase at that time. Instructors were still being released from compulsory wartime service, and many were simply waiting out their time and counting their days to being demobilised. I joined them. There were also more than a few resentful sergeants with B.Sc.s doing National Service in the ranks rather than as the gentlemen that they had hoped to be. The teaching standards, therefore, tended to be somewhat variable, to put it mildly, and one of my early officer teachers for example, who hailed from Canada and was the soul of goodwill, managed to teach us all the words of 'Eskimo Nell' in one term, and attempted absolutely nothing else at all. He really was poor and untalented at his profession, I now realise, because I can only remember some of the verses, and particularly not the better of the nice vulgar ones.

One of the band of resentful sergeants named Zahler, was taking his duties very seriously indeed. He had obviously been warned that we apprentices might on occasion be fractious and difficult to deal with, and was accordingly taking the elementary precaution of carrying a knife in his sock for protection, and at regular intervals showing it to us and reminding us that he was quite prepared to use it if we were foolish enough to start anything with him. It was most flattering. Apart from these shall we say small teacher/student psychological problems, at this early stage of our technical education there were no serious difficulties for those of us with some existing education in engineering, but for me and some others it made for a very uphill struggle, and even decent teachers of whom there were few, tended to focus their attention on those who already understand anyway. My own particular fall-back position was to observe and try to learn at least something from 'Horace'.

The electricians were doubly unfortunate, as they had the squadron leader head of the electrical department as their teacher for engineering in the first year. This was an officer who enjoyed an immense natural flair for bureaucracy, and his true interests all too discernibly lay in that distant magic day when he would deservedly inherit ultimate power, become the headmaster and give up all this kind of nonsense, and in the meantime he visited his frustrations upon us by doing his level best to enjoy the smaller joys of socking it to us as hard as he possibly could on every possible occasion which presented itself, and there were many of those for a man with such inclinations I can assure you. His level of concern for the care and education of the young and impressionable had plainly been inherited from a very long and ancient lineage, touching on Charles Dickens's Mr Squeers, and on his mother's side was directly traceable back to Attila the Hun, I believe. My own mechanical engineering education went into mental suspension the day that this officer announced to me irritably in answer to my perfectly legitimate question, "It must be obvious, fool, that a push is the same as a pull." Unfortunately it wasn't, and to this day it still isn't, and whilst I sat back in shock, considering the deep ramifications of this obviously vital piece of groundwork, I of course missed out on the next big chunk of essential information as well. Even my good friend 'Horace', with the best will in the world, could be of no help to me on occasions such as these.

The school's upper establishment contained more than just one of this kind of gentleman. Apprentices were allowed a mug full of milk as part of their daily ration scale, and, as an added bonus, the sheer luxury of an accompanying rock cake which was served to those who queued for it at their morning break times. Although permanently hungry, of course, many of us had no great liking for either queuing or milk, feeling that the latter tended to too closely identify one with rather recent juvenile activities. Additionally, the carrying of the mug entailed an element of risk that was more than likely to produce a breakage before too long, and therefore give rise to the immediate need to purchase a replacement from ever slender resources. On balance, the milk was not considered to be worth the effort by the majority of us. The deputy head of schools issued an edict that, in future at schools, the miserable rock cake would in no circumstances be issued unless the apprentice presenting himself also had a mug full of milk to show. Definite shades of having mistakenly studied *Oliver Twist* as a recommended education method guide in his time, you must agree. What is more, he voluntarily gave up his own break times in order to enjoy prowling amongst the groups of standing apprentices, trying to catch those vandals amongst us who might jettison the milk in the nearest bush, after using it to obtain the rock cake. The result was that, like some criminal about to be arrested, one stood to rigid attention, and held out and displayed empty hands if he chanced to put in a sudden appearance amongst you, and you just learned to go hungry.

HARDENING AND TEMPERING

The village blacksmith, he was there

In spite of my evident lack of ability to be everything that the Air Force longed and hoped for in the wing area and my struggling approach to schools, I still held hopes that I might after all prove to have some talent for learning to be a decent tradesman at workshops. However, my first look at the interior of the electrical workshops indicated that even here all might not be entirely well either. Unlike the airframe fitters who enjoyed newly constructed spacious and well planned hangers for their classrooms, our buildings were thoroughly circa 1920s factory in character, damp brick squalor, concrete floors and all, and had been constructed by German P.O.W.s at the end of the First World War. The long bleak bays were divided into classrooms on each side of a central walking area by simple freestanding wooden partitions in a tasteful unvarying dull blue, and rough wooden combined desk tops and benches of the same colour were the sole furniture. It was all rather primitive and it tended to show.

After we had been equipped with overalls and tool kits, our workshops training began with some several months of learning the arts of basic fitting, an experience that I had not taken into consideration at all when I had elected to become an electrician. The Air Force had still not quite caught up with itself at this point in time, and was still trying to turn us out as craftsmen in the old pre-war image rather than the service by replacement tradesmen that it would eventually happily settle for. Plainly we were intended for something better, but fourteen years later I was still none the wiser about what it might have been. Basic fitting included the skills of using hand tools such as hacksaws, files, drills, taps and dies for openers, before moving on to lathe turning, blacksmithing, coppersmithing, silversmithing and the like, and yet again was completely outside my world of experience. Paper folding and flower arranging would have been just about within my hand skills at that time.

The very first task for any self-respecting apprentice, however, was to drill a small hole through the centre of his brass four-bladed propeller arm badge and tap a thread into it. The badge could then be secured with a brass screw, and locked in position by a halfpenny, similarly drilled and tapped and located inside the jacket sleeve. It was a matter of individual pride that you did this yourself, and that, when polished and finished, it was impossible to detect with the naked eye just how the wheel was secured. Whilst the spirit was willing enough my flesh was not yet very skilful, and my lack of experience caused me to select and use a screw a shade too long, which inflicted some discomfort to myself when shortly afterwards on a drill session, I was slapped on the very spot with a pace stick by the regiment sergeant, waving it around wildly in one of his more exuberant moods, but one had to be philosophical and, as they say, there is no gain without pain.

In the wartime school system which I had enjoyed, one was fortunate to receive any wood or metalwork crafts education at all due to the absence of suitable teachers, who were in the forces. As a consequence my hand skills in metalwork were completely non-existent rather than merely deficient as I have said, and again those with a polytechnic background had a distinct advantage.

The electricians had been divided into two classes for ease of handling, and our 'A' class instructor was a civilian called Mr Tripp, himself a former boy mechanic circa 1924, whose instructional technique was to say the least something to marvel at. Without going into detail it was broadly to gather us around him in a group, tallest and fattest at the front, shortest at the back where they could not possibly see, and demonstrate just once by doing it himself, then give out the materials needed, and depart for the more convivial surroundings of the instructors' rest room before anyone could ask some wretched question, leaving us to manage as best we could. It was very much as follows:.

"Saw it in half with the hacksaw like this, see?" Sniff, followed by two quick strokes of the hacksaw. "Then file it flat to the right size like this see?" Second sniff, regulation two passes of the file. "And leave it a bit bigger so you can put a good final finish on it with your smoothing file. No questions? Good. The materials are on my desk," Final sniff and exit stage right seeking the solace of his fellow civvy instructors. Of course the newcomers to hand tools like files and hacksaws, with soft young hands steadily fell behind the more experienced, as they, with the advantage of some existing skills, naturally progressed more speedily to the next task. At that time Tripp would suddenly reappear amongst us like the demon king in pantomime, and then simply demonstrate the latest new job to the whole class. On occasions I found myself as many as three whole jobs behind, yet being taught, if you can call this teaching, the method and skill for the task of the fastest and best fitter member of our class, who was Terry Thornton, by the way. My first most urgent matter was to try and persuade Terry to proceed at a much more pedestrian pace in the interests of my particular survival. Even so, by the time that I got to any particular task I couldn't remember exactly what it was that I had seen, or its special significance, let alone develop the skills. Without help I made no progress at all, and blundered into every imaginable mistake. Tripp's view seemed to be that this was just hard luck on me.

I remember that one early exercise was to make a two piece hinge from a thick brass plate. This meant a lot of hacksawing and filing, and it all had to be within a thousandth part of an inch. I simply could not yet file it to a flat surface, let alone to any specific measurement. My final version was neither square nor flat and to no particular size at all. Whilst providing some humour for my classmates when Tripp held it aloft and showed it to all and everyone, it angered me that he had not even been present for most of the many hours spent on attempting it, let alone seeing my problems and helping me.

A further problem was that some of the tasks were to manufacture your own tools for future work, such as making your own lathe tools in blacksmithing for example. If you chanced not to make good ones, at a time when you did not yet even know what a good one might be, then later your lathe turning work was likely to prove to be even more difficult. The world was full of surprises and the blacksmithing instructor advised us one day that we should pee in the water trough used for quenching the hot metal. At first I thought that this surely must be one of those unfortunate jokes applied to young beginners such as ourselves, and that the moment any one of us as much as started to unbutton his trouser flies something very nasty indeed would happen, but I was quite wrong and there is a scientific basis in fact for this being good practice. This does not imply in any way that dog pee hardens lamp posts by the way. There was a poster on the wall of the blacksmith's workshop that I recall which advised:

> *A chisel u/s?*
> *Repairable true*
> *So Bill at his forge*
> *Reshapes it as new*

There was nothing even in the small print, however, to indicate if our good friend Bill urinated in the water trough or not.

I still have some of those tools which I made at that time, but unfortunately not the brass hinge which I would now deeply treasure, because Tripp refused to let me have that particular item when we finally said our last farewells. Difficult to the very last, he let each of us choose only three of the tools that we had made in our time there. There was a glass case containing the best tools ever made by electrical apprentices and Terry eventually had several exhibits in there. Indoors or out, Tripp always wore a cloth cap and no one had ever seen him without it on, even to scratch his head. An entry previous to mine had visited workshops to say their farewells after graduating and, being light-hearted, some of them had removed this cap as a joke. He still reported them, and on arrival at their next camp they were immediately put on punishment.

The electricians who had been grouped into the second of the two classes, the 'B' class, fared little better than ourselves, and had a rather short-tempered corporal, whose name I cannot recall but whose manner I can only too well, teaching them basic fitting. Pat Cropley found himself in deep trouble once when he could not find the correct handle to get his lathe out of automatic feed, and called out too late for help. His reward was to be put on a charge for wilful damage. One of my classmates, Robin Berry, was still so small that he could barely reach the top of the workbench, let alone file on it, but our instructor ignored this. It wasn't his problem, it seemed. The rest of us arranged our wooden tool boxes to give Robin a bit of a leg up and a base to work on.

In the meantime, before we progressed as far as any of the electrical subjects in which I was really interested, I avidly read the many posters which decorated and covered the worst of the damp on the brick walls of the old workshops, seeking advanced inspiration on how to master the electrician's art. Most were dire and cryptic warnings often in glorious Technicolor such as, 'Wires, Pliers, Flashes, Ashes', and, 'You'll End up in a Wooden Box, if you Jamb the Interlocks', was another somewhat gloomy prediction that I recall, although, in fourteen years of service, I must add that I never as much as found a single interlock to attempt to jamb. Another was, 'Many men have ceased to be because they touched the E.H.T'. Although I wasn't quite sure what E.H.T might be as yet, I vowed to keep my little fingers well away from it since I was quite interested in survival. Yet another began with the cartoon tale of 'A.C. Duff whose motto was, I treat 'em rough, and A.C. Bright whose motto was, I treat 'em right', and their individual differing approaches to servicing the R.A.F.'s magnificent battery electrical equipment. Although Bright obviously had the better thoroughly responsible service attitude, something in Duff's easy-going, light-hearted, I-don't-give-a-monkey's approach and manner always seemed to appeal to some sympathetic chord within me somehow.

The happy day arrived when we of the 'A' class had finally completed our first spell of basic fitting, and departed from Mr Tripp's loving care to that of masters new. One of our early workshop class subjects was field telephones and, for some practical work, we were finally given the task of setting up the portable telephone exchange at the end of one workshop bay, and groups of two or three then ran out drums of cable to various far points of their own choosing, connected up their hand sets and got the whole enterprise up and running. There were remarkably few places to set up these hand set outposts however. Anywhere even remotely close to some other ongoing class or other only resulted in being chastised by the instructor in charge, and told to go elsewhere in less than polite phraseology. Accordingly one group were obliged to locate themselves in the relative peace and quiet immediately below the instructors' offices, and there they made contact with the exchange two whole bays away, where 'Swing' Swoffer had wriggled himself into the operator's chair, watched by several colleagues, myself included, anxiously waiting our turn. The dialogue was naturally rather coarse and noisy which eventually aroused the interests of Warrant Officer Smith, the then senior electrical supervisor, whose office window happened to open just above the giggling noisy pair.

He came down the steps angrily and snatched up the hand set from their fingers before enquiring into the mouthpiece, "Hello! Hello!"

Swoffer, believing that this was just some extra piece of ongoing light banter, and having recently taken up swearing in depth, then happily began to go through his latest full repertoire of vulgar and disgusting expletives with our giggling background to support him. The enraged tones which resulted convinced him that he had perhaps made an unfortunate connection, and he hurriedly snatched off the headphones, and handed over the exchange to the next one waiting with the explanation, "Here, it's your turn. I'm off to the bogs."

Meanwhile Mr. Smith was following the cables along, literally hand over hand, tracing them back to where the exchange might be located. It was only the thoughtful actions of the pair below the window, who waited until the red-faced warrant officer had turned the corner of the bay, before issuing their warning to the rest of us back at the exchange, that saved the day. We took their good advice and joined 'Swing' sitting in his place of safety with the door closed, and gave him a good verbal going-over for almost dropping us all in it.

Throughout my time at Halton, I was obliged to periodically volunteer for extra technical training and give up my sports afternoon, solely to attempt to overcome this basic fitting problem. The teaching method didn't change in the least during these extra training periods either, when it would have been entirely possible to have a more one to one type of lesson. Our civilian friend Tripp simply ignored me and disappeared as usual, leaving me to teach myself and that is exactly what I eventually did. Once our class had progressed away from basic fitting and on to electrics, however, I came far more into my own and began to do much better, and I was placed first in my entries electricians at the end of our first year examinations. When one of my classmates, with his tongue in his cheek, and wishing to indulge in a spot of bar rattling, told Tripp of my small success on our return to his tender mercies for a second spell of basic fitting, his reply was that he didn't understand why I was so stupid in basic. I in turn replied that I wasn't, just badly taught. That brought me a strong reprimand from the officer in charge of electrical workshops, and I was lucky to get away so lightly, what is more, despite the provocation.

In fact this was very much my story throughout early training. I struggled at schools initially, through lack of a previous technical background more than any lack of ability until, in the final year, we electricians at last got a really good teacher, and, what is more, one who wanted to do the job well. All of us owed an awful lot to the way that he gave up lots of his spare time to privately coach us in the period leading up to our school finals, and try and put right what had just been allowed to go sour by neglect over the previous two years, which only went to show what might have been achieved. At workshops, apart from basic fitting, I was never lower than the top three or four in my entry, but it took hours of extra study on my own to achieve it. I was not naturally talented, as was 'Dagwood' Large for example, and without doubt he was easily the best electrician from our entry. I had to work much harder for anything that I managed to achieve. I was always so much better at things which required my head rather than things which required my hands, exactly as I had realised before joining the R.A.F., but unfortunately I was in the wrong Air Force for that, of course, and no longer in any position to do anything about it either.

We were now all living three separate lives each with its own demands to adjust to, and in the Air Force's eyes each was apparently equally significant in the development of their desired final product. In the wing I was not prepared to creep, and was inclined to get on the wrong side of authority when it was unjust, which was most of the time. I found it rather hard to accept that, as one of our disciplinary people once put it, "An Hairman his a higorant hanimal hand should be treated has such."

In particular I did not enjoy the times when we were delivered into the hands of the regiment. My interests and efforts at that time lay in trying to become a good tradesman, not in crawling in the mud to some better way. If that is where their instincts took them, then that was their business and probably appropriate. Personally I found them dumb and despicable at that time. It is worth noting how this conflicted with what they thought of us. When we finally graduated, only one of the six electricians who passed out in the higher grade of aircraftsman 1 had been promoted to any rank of N.C.O. apprentice, but from the Fifty-Fifth electricians came one sergeant apprentice, two corporal apprentices and four leading apprentices. Some of my companions were obviously experiencing considerably less difficulty in adjusting to these values and this kind of life than I did, and still seemed able to regard it all as a jolly kind of boarding school with even an opportunity for a spot of bullying thrown in as soon as there were entries junior to ourselves, but already it was no longer anything like that for me.

BLESSED ARE THE Bs
FOR THEY SHALL EAT FIRST

This old coat of mine has had its...

Playing football outside our barrack blocks one sunny Sunday that first summer, I noticed that I seemed to be developing a bright red rash which did not go away with the application of soap and water in the shower, and by evening I was in the Princess Mary R.A.F. Hospital just down the road from Maitland wing, suffering from what was diagnosed as being German measles. There I remained, now dressed in R.A.F. hospital blue, which by the way fitted me far better than my normal uniform, until I was no longer considered contagious, and what is more, I was well fed and living in comparative ease and comfort, but, more fool that I was, I somehow missed being part of all that daily toil, and I agitated daily to be allowed to return to my new comrades.

When I did finally return to the companionship of block 7 room 5, I discovered that as a normal procedure all of my kit had been bundled up in my kitbag, taken to some distant place and immersed in high temperature steam before being stored to await my return. This may have gone some way to preventing any subsequent spread of disease, but an unfortunate practical side effect was that my greatcoat was now faded to more of a wehrmacht field grey entirely in keeping with my disease rather than an airforce blue, and possessed creases in many unusual places which no amount of ironing and pressing on my part managed to eliminate. I awaited the coming autumn with the certain expectancy that I was in for some trouble. All the rest of that summer, as the garment lay in my locker tightly folded in its metal former with polished buttons showing, it passed unnoticed, but, as the autumn came around, it was time to wear this item of clothing again, and so one day I chanced to be one of the last out of the barrack room and forming up in ranks with the others of my entry, dressed for the oncoming winter. Never slow to notice something unusual, a shout went up from one wag as my Joseph's greatcoat of differing shade came into focus. "Heil Von Grim!" Everyone enjoyed the humour, and, being the kind of apprentice that I was rapidly becoming, I inevitably became instantly far more deeply attached to the garment. I was not to keep it for long, however, the regiment man saw to that, although the fault was not mine, and I had to bear the cost yet again from my miserable pay. By the time that one was supposedly to receive one's accumulations of unspent lucre, I found myself to be in deep debt for my share of one demolished N.A.A.F.I. building, one best uniform that I had failed to grow into, and one greatcoat driven prematurely grey by its short association with me. Fiscal matters tended to continue in this fashion and I never did quite manage to achieve a surplus in my account the whole of my time at Halton.

All that first summer too we were all being inoculated and injected against any and all diseases known to medical science at that time, and a few previously found only in parrots. The most commonplace and boring event of the day was some poor devil fainting away on the parade ground each morning either immediately after or prior to yet another jab or scratch. Vaccination laid us out by the tens rather than the singles, myself included, I recall. Some, in the act of falling, bit the next one on the way down in a tender place, and then he went down too in a domino effect. I, like many others, was not an entirely well man that first season I recall, but it excused all of us absolutely nothing under a general toughening up regime obviously intended to create us in the lean and mean mode.

Sports afternoon and my love of football led me to becoming a referee, but only after a searching interview and extremely detailed questioning about the rules of the game by the station sports officer, himself a well-known international athlete, and I was soon spending my sports afternoons officiating at some of the better matches. My first was a station cup tie between an apprentices' team and an airmen's team, and the aforesaid station sports officer came personally to watch how I made out, probably realising too well the propensity for some very real violence with the relationship between the apprentices and the other camp personnel being what it was. He was not to be disappointed, and, put it at its absolute mildest, it was what could be described as a very rough match indeed. Within the first quarter of an hour I had to send off one player for a foul of plainly hospitalising intent on the opposing goalkeeper, which would doubtless get him a regular first team place in several Premier Division professional sides these days. At half-time, as I sat alone at the side of the pitch trying my best to demonstrate my lack of partisanship despite my apprentice's wheel arm badge, the sports officer strolled over and gave me a piece of good advice, which I have since found applies equally well to management in general as to football in this kind of situation. He told me, "Now that they seem to have got the general idea, only the obvious from now on."

Although he and I might well have agreed on this point, no one had apparently told the two teams who, on the resumption of the game, went straight back at it hammer and tongs. No prisoners were being taken on either side that day, and when I awarded a corner to the apprentice team later in the game, the airmen's goalkeeper shouted out, "Why don't you give them a penalty, ref?" His name went into my book faster than the proverbial through a goose, and the sports officer hastily departed, presumably not wishing to witness my getting around to abandoning the game altogether, which appeared to be the most likely outcome at the time. My written report on the game covered almost two foolscap pages.

The gymnasium was located at number one and two wing, and was where we first had our induction medical tests and interviews. It was absolutely magnificently equipped for all indoor sports, and the outside sports fields for soccer, cricket, rugby and hockey stretched as far as the eye could see. The sheer availability of sports of all kinds at Halton was simply breathtaking and could not be faulted in any respect. For the very first time I saw and enjoyed a wide range of sports quite new to me including basketball and fencing, and watched inter-entry boxing matches, in one of which by the way my old school friend Jack Gough got himself knocked out in the second round by a certain Sergeant Apprentice Knapper, from the Forty-Seventh as I recall. At least they both had been of approximately the same physical size, but at a later time I witnessed yet another inter entry boxing bout involving a member of my entry, who although quite brave and not entirely without skill, was quite small. By some gross miscalculation, he had been drawn to fight a member of a senior entry who was, to put it mildly, massively built, all of six feet three inches tall and who enjoyed a somewhat unhealthy reputation for violence, whether under Marquess of Queensbury rules or otherwise. He had as a matter of fact put his previous opponent in such a bout in hospital for a protracted period. The Flight Sergeant P.T. instructor in charge of our entry's boxing team was none other than our good friend 'The Muscle Man,' who had quite nonchalantly informed my dismayed comrade exactly who was to be his next opponent, and advised that, as well as keeping constantly on the move if he wished to avoid the same unfortunate fate as the previous opponent, he would most certainly need a good sound game plan. As a piece of encouragement he had added that the massive opponent was well known to have a glass jaw. The reliability of the source of this piece of information or just how my comrade was supposed to reach this point of apparent weakness was not revealed however. When they were called to the centre of the ring by the referee for instructions, Goliath was already punching the air violently, rolling his eyes and making quite shocking animal-like growling noises. My comrade was not long into the bout before bringing his game plan into brisk and urgent action. He embarked by lashing out with a blow which, had it remained at its lowest point, would have effectively removed both of his opponent's knee caps, but was however suitably aimed to be slightly more tangentially upwards in direction than that, and suffice it to say that the intent was so clear that he got himself immediately disqualified and sent to his corner by an irate referee, but my comrade was generously applauded and was very much a hero, at least in the eyes of his own entry that night.

Our education stretched over a very wide area indeed, including the spiritual. Nothing at all was being allowed to grow unattended, you see. The regiment sergeant was dividing us all up by religion one day for some strange reason that I no longer recall, although it certainly was a sports parade I remember. Already blank acceptance was rapidly replacing fruitless logic, and such a concept as this did not unduly alarm me any longer. He was probably just demonstrating yet again that, whilst we all might be equal under the eyes of God, we most certainly were not under those of the Royal Air Force. This occasion was something a little more serious than the customary morning parade order of, "Fall out the Roman Catholics and Jews", which was intended to spare them hearing what we religious mainstream at the back could not hear either. Our entry had accordingly adopted a common negligent policy so that no one at all bothered to fall out, and the padre, on noticing this, kindly refrained from anything which exceeded the general guidelines of who would eventually inherit what.

To return to Regiment 'Jim' however. First he had all the Roman Catholics form themselves up as a separate group at the rear, then the Jews immediately behind them and so on, then finally the Church of England main body. To the sergeant's surprise, this still left one of our armourers, 'Donga' Vennant, in splendid isolation in the original position. He went up to him and bellowed, "Vennant! What doo yew think thart eet ees thart yoo are dooin boy? Are yoo tryin to take the pees out of me or sometheeng?" Don replied, as was the truth, that he was a Buddhist, but back came the reply, "Then why haven't yoo moved back with thee other Jews then lard?"

Another of his early jocular favourites was to insist that we wait for the actual word of command and not anticipate it. He would then continue, "When I shout worn yew weel orl jump in the air, and when I shout two yew weel come down again." He would then shout one, and of course find some small humiliating punishment for the first one whose feet happened to be seen hitting the tarmac.

Once a fortnight we were paid, and, as you might well guess, that also entailed a special parade which was conveniently arranged to be in our time rather than theirs, namely, in the lunch hour preceding our sports afternoon. When we had been marched back from our morning's labours to the wing area, we would be assembled by entry, we the Fifty-Fifth entry outside in the road regardless of weather, to which we were now considered to be totally impervious, and the Fiftieth inside the 'tank' for this delightful purpose, thus ensuring that half our lunch time was already gone before the Fiftieth entry had all received their pittance, and we could now be moved inside in turn to receive ours in strict alphabetical order. The ritual was that the pay clerk would call out your name, to which you responded by coming smartly to attention, calling out the last three digits of your service number and marching forward and saluting the pay officer. You would be handed your ten shilling note, then you turned smartly right and marched away. All this for half a bloody quid – I ask you. I thanked God that in his wisdom he had seen fit to make me a Bristow rather than a Wallace because, as a 'B' I at least stood a fighting chance of snatching a mouthful of food in the time remaining to me before we were back on parade for the march en masse to the next piece of planned evil. Poor 'Horace' had rather more limited options open to him. It was either go without or attempt to slip away immediately to eat before the first ones of the Fiftieth entry pushed in front of him, and then hope to slip back in our ranks undetected before the 'Ws' were called out. Many years later, the nostalgia of all this happiness was revived for me one day when I was myself detailed as the pay parade sergeant at a training camp. Among the airmen to be paid that day was a fellow who, shall we say, was known to be somewhat eccentric. When his name was eventually called out he shuffled forward, muttered his last three digits, but did not salute and waited for his money with outstretched hand. The paying officer that day was a young inexperienced national service man and he huffily asked in his little piping voice, "I say! Don't you salute a commissioned officer then airman?" To which the eccentric one replied, "Jim, for another quid you can have an effing march past if you like."

I needed every ounce of my self-control to not show any smile that day. He must have been an ex-apprentice driven to this condition by our common earlier experience of pay parades, I speculated.

With the financial situation as it was, and since I had recently taken up smoking as a palliative to my now frequently frayed nerves, albeit only five Woodbines a day, it had become increasingly necessary to at least try to obtain some form of supplementary income, and fellow electrician 'Chiv' Chivers and I combined our forces. He possessed an ancient pair of rusty hand clippers, and, with no previous experience other than sitting in the barber's chair between the two of us, we attempted to go into the hairdressing and masculine coiffeur trade. With so many suspicious people about, it was difficult to find someone willing to be our first practice model, even at our bargain basement prices, but we did in fact eventually locate one the evening before a Saturday parade. It was an engine fitter named 'Blondie' Wilson who was in a desperate plight, in that he had failed to get the job done at the camp barber's after being ordered, and therefore a fate worse than death awaited him the following morning with absolute certitude. I cannot claim that he was exactly enthusiastic about the whole business, but after all, needs must when the devil drives.

I began after an after-you-Mr Chivers, no, after-you-Mr Bristow brief preliminary. It wasn't too long before my hand became terrible tired from the stiff clippers, and Chiv took over and he in turn had to give it up as well before too much progress had been made. We finally compromised by agreeing to at least continue until 'Blondie's' hat would conceal the true state of underlying affairs, and to 'put a finish on it' as our basic instructor Mr. Tripp used to say, at a later date. Our client was not amused, but he was the eventual possessor of the very first Mohican cut, and years ahead of his time, if he had only had the wit to see it in that light. Without a satisfied client to advertise our obvious abilities, the clippers languished in Chiv's foot locker, as did my trumpet mouthpiece, for want of another customer for the remainder of our stay at Halton.

I was not the only one to have taken up smoking on slender resources, and cigarettes became a luxury item which like everything else had to be shared. This involved a new language of its very own too, and a call of, "Twos up on your dimp," or, "How's about a quick draw on your spit," at some break time or other indicated that the nicotine-starved speaker was desirous of sharing a few puffs of the fag end of someone's cigarette.

NATURE AND THE GREAT OUTDOORS

We'd be all right in the middle of the night
Whatever the kind of weather

Another joy awaiting us was summer camp. This was the time when we were delivered full-time into the tender mercies of the R.A.F. regiment, and taken away from normal training to live under canvas for a couple of weeks. This was a part of general service training, but in fact was more what I would call playing soldiers. The first year, the site chosen for this event was relatively close to Halton and we marched there in full kit one Sunday morning. It had an auspicious beginning when our old friend the sergeant, marching along beside us with nothing more burdensome than his pace stick, decided to brighten up our obvious gloom and ordered us to sing. The response was not encouraging, just a plodding silence. He then began to be more assertive and ordered, "Sing or double." Had he used our own variant of this, he would undoubtedly have seen two hundred naked bottoms at once. As one, we broke into double time and that seemed to set the mood for the whole of our time at the camp. It was long route marches, day attacks, night attacks and all that sort of awful thing.

We were aroused late one night from our tents after yet another exhausting day and made to stand outside in the dark listening to noises which we were required to identify. To help focus your mind and to motivate you, it was made clear that, until you did correctly interpret the mysterious sounds, you were not going to return to the comfort of your bed. As desperate failure followed desperate failure, the guessing became more and more wild, and, considering that a typical sound selected for our ears was eventually revealed as being that of a sentry urinating on leaves, the prognosis for sleep was not good and I was one of many who were still standing there when the dawn came up. Personally, I have always felt that it is not a sign of an entirely healthy mind to display an overly active interest in the sounds of others performing their bodily functions in the dark.

To further my enjoyment of the situation and more quickly adjust me to my new circumstances, the regiment arranged that, when the sun was just a shade higher, I found myself crawling through all kinds of nature at its most odious in some woods, a member of a section seeking an invisible enemy, with my friend the sergeant providing a running commentary and kindly advice.

"Yew," he said, putting his hob-nailed boot firmly on to the back of the hapless creature selected to be section leader and pressing him firmly into the ooze. "Yew are dayd, shot by a sniper. Number Two take over!"

We had not proceeded very much further before it was announced that poor number two had apparently suffered a fate similar to that of his unfortunate leader, and it then became my turn to lead. Acutely aware that some positive action on my part was expected, I made the appropriate hand signal, turned the whole section round and began a slow crawl, back in the direction from which we had recently come. He was there in an instant, the hob-nailed boots poised inches away from me. "Arnd what doo yew theenk yew are dewing?" he queried.

I replied as courteously as was possible in the circumstances, "It's extremely dangerous out there sergeant. In fact people are being killed."

His withering look indicated that I would probably be receiving some further instruction in the immediate near future.

The following day, when we had all been divided into small classes, I was yet again so fortunate as to be selected as a member of a group receiving our morning's instruction at our good friend's tender hands. We began with a long march to some long-forgotten outpost of the empire just to warm us all up, and then it began again at its muscle-numbing worst. As a direct result of the long return march our group was exceedingly late returning to the camp area for our midday meal. A pure bonus to him, of course. When we did finally arrive in the proximity of our cosy tented accommodation we could see the rest of our entry in plain sight already gloomily queuing for their food outside the mess tent. It wasn't quite over for us yet, however, and our friend halted us and informed us that, as a final pleasurable act of the morning, we would now fix our bayonets and carry out a charge on our hungry comrades, blood-curdling shouts and all. To try to raise some shred of enthusiasm among us, his voice grew louder and his bright eyes began to glance at us all in turn.

"Yeew weel," he began, "Yeew weel sweep therm orn worn side like an," and here he paused, seeking the absolutely correctly inspiring phrase before continuing. "Like an irresponsible force," he shouted and he punctuated it all by making a sabre-slashing sweep of his right arm horizontally in front of him. Well, he had certainly managed to get one thing right at last, I thought. We reluctantly rushed forward on his shouted command only to find out precisely what happens when an 'irresponsible force' meets an immovable object, particularly if the immovable force has itself not been enjoying a morning entirely blessed with smiles and roses, and is plainly none too happy about it either. For the benefit of those without such experience, I can tell you that the immovable object stays exactly where it is, bidding the 'irresponsible force' to "f*** off for Christ's sake," and the 'irresponsible force' recovers its knife, fork and spoon from its hidden pockets and joins the end of the queue.

Another convivial evening was spent by us all, attempting to extinguish a hurricane lamp situated in the middle of a group of defenders, spaced widely around it. The apparent objective was to avoid detection whilst crawling into the ring in the darkness, blow out the lamp and crawl out again still undiscovered. Pat Cropley and I, both sharing a common fear of naked flame, lay in the security of our little bush, and slept until the whole regrettable procedure was over.

Our good friend had been giving us a lesson in fieldcraft one day and impressing upon us the need to be precise, and to use commonly understood terms when indicating the location of some object to another. We had been made aware of such subtleties as 'red roofed house, two fingers left, bushy-topped tree,' and the like, rather than, 'a bit left of that charming old gabled semi-detached there seems to be an old elm if you care to give it a momentary glance.' There was something special in the sound of 'bushy-topped tree,' that seemed to particularly appeal to our sergeant friend in some special way although as far as he was concerned there were only three types, bushy topped, pine or fir. Probably this preference dated from a childhood association when he had fallen on his head from one of that variety, and his closing remarks were to the effect that this particular phrase should be borne in mind at all times. He then jumped to his feet, glanced up at the sun, remembered from its acute position that it was high time to play the bastard again, and called us all to attention. He then began a brisk inspection of our weapons, whilst giving himself time to think of something even worse. Something of interest seemed to catch his eye in poor 'Horace's' rifle barrel, and our friend beamed in the joy of anticipation of dishing out some punishment, but naturally, first he was going to inflict a degree of humiliation on the individual in front of the rest of us, as a warning to the wise.

"What do yew theenk I have found ere?" he began to us all with a smirk.

As one voice the answer came back from the rows of smiling pupils.

"Bushy-topped tree," they chorused happily.

This did nothing to alleviate any further suffering for poor 'Horace' of course, quite the contrary. Many years later the subject of bushy-topped trees was to again crop up between them, as I will later recount.

The sergeant and his friends had been teaching the art of camouflage on another day, and he had become somewhat boastful of his prowess in this field. No one could detect snipers whom he had hidden, it seemed, and to prove it here and now, before our very eyes, mark you, he would hide no fewer than six, yes I say six, men, no further than a few yards from a main path, and the rest of the sections would patrol down there for a half-mile or so and not see a thing. I was one of those selected to be the unfortunate gang of six. He chose to hide me close to a large bush partly covered by a net, and liberally rubbed my face and hands with the evil-smelling glutinous mud to conceal any white areas. There was to be no wimpy camouflage cream for us. The others received no better treatment, some even being obliged to lie motionless in ditches half-full of stagnant water. When all was ready to his satisfaction, the sections were started back down the path one at a time. I had managed to absorb sufficient of our friend's lesson to realise that the whites of my eyes were probably the only white area that might be detectable, and, as the first section came round the bend and into view, I began to frantically blink my wide open eyes. Any kind of movement larger than this would inevitably lead to his detecting it and my punishment, I realised. I believe that it was my fellow electrician 'George' Hanchett who spotted my subtle alarm, and what is more, understood my intention, because rather than calling out the required "Down!" alarm command, which would have brought the whole section down prone on the ground in an instant, he stood where he was, pointed his finger at me and shouted in obvious joy, "Look! It's bloody Grim." The lesson was abandoned and 'George' and 'Grim' shared the guard duty that evening.

The weather was not kindly that summer as it had not been in the preceding winter, and it grew steadily worse while we were at camp. The unremitting rain did not initially disturb the daily and nightly programmes, but, as the tent areas became a quagmire, it became impossible to keep oneself dry or to dry already sodden kit. When it was so bad that even 'they' began to notice it, our Wing Commander arrived for a visit unexpectedly one day. He had us all assembled in the mess tent, stood on a table and made his speech. The content, as I recall, was that he was proud of us, nay extremely proud of us and the way that we had endured the conditions, but things were now so bad that it might be better to terminate training temporarily and return when the weather was less inclement. In an act of uniquely uncharacteristic democracy, the choice would apparently be left to us. There was the most brief of possible silences then began a terrible cat wailing plea from two hundred semi-bronchitic voices.

"Go back! Go back!" the chant continued. He gazed around us from the prominence of the table and finally said, with outstretched arms and tear-wet eyes, "Very well apprentices, you may stay,"

Then he got down from the table and departed to a vast unbelieving silence.

The weather did not improve, I failed to become a devotee of the sounds made by sentries in the dark, and my humanitarian concern for the safety of my fellow creatures' skins seemed to have brought me to the closer attentions of our good friend the sergeant. We finally returned to the comparative luxury of our barrack room and the tender disciplines of wing life, dragging our muddy possessions behind us.

WINNING NEW FRIENDS

When he chanced to spy
Poor little Angeline

Our wing life mentors and terrorisers were by no means infallible, we soon learned. Sergeant Dick Corser, our disciplinary overseer, had decided, he told us, that he was going to concentrate his full energies on teaching us his particular individual version of drill, and making the Fifty-Fifth entry a living example of the very highest form of that art and a wonder fit for the whole world, including any Guards regiment, to behold. Some people just settle for keeping pigeons or collecting stamps, mark you, but there you are. His concepts were that there should be a marked pause 'equal to two beats in quick time' as he put it, between separate parts of drill manoeuvres, and soon he had us all shouting out this "left right" rather than the usual 'one pause two' thing as we carried out his evil bidding. On reflection he had probably picked up this idea from the advanced drill squad who frequently used our parade square for their own practice at that time, and who had a lowly airman in their rear rank with the unenviable task of mournfully calling out the time of "Eff, Oight" for all the others.
    We all got quite good at it too, I remember. Until the day of the Forty-Seventh entry's passing out parade rehearsal, and our very first appearance at such a grand event as this, that is. We as junior entry marched on last, at the rear of all the junior entries already in position, halted and waited for the sergeant apprentice in charge of all the junior entries to order us all to left turn into line together. When he finally shouted the order, fifteen hundred pairs of boots hit the ground together as one, followed after a short pause 'equal to two beats in quick time' by another two hundred. The Fifty-Fourth entry immediately in front of us turned and sneered and muttered 'shower'. It was a most humiliating debut. The matter of the pause was thankfully dropped by authority soon after that, but that 'shower'

retort was neither forgotten nor forgiven, and was stored away for a future day that was in fact coming sooner than we thought.

A typical example of the kind of actual bodily harm that could befall you by chance in the wing occurred one Saturday morning, when again we were all rehearsing some entry or other's passing out parade, I recall. Every entry was on number one wing's parade ground, drawn up in long ranks with senior entries at the front and junior at the rear as usual. All were dressed in full white ceremonial gear with rifles and fixed bayonets. It was a very impressive sight indeed in those days of large entries, with almost two thousand apprentices out there. Of course, we from number three wing had started much earlier than the rest because we had the furthest to march, and as junior entry of that wing we had waited the longest before marching on. It was a blistering hot day and we stood there at attention for a very long time. One of our entry, still not sufficiently hardened, fainted, and as he fell the bayonet slipped under his arm and impaled his shoulder. He lay on the ground in a growing pool of his own blood and of course the boy next to him immediately started to move to help him. The regiment N.C.O. behind them immediately ordered him to stand where he was and leave him, and there he remained until it was all over.

Just sometimes fate seemed to take pity on us and deliver into our hands a piece of light entertainment. One summer's day while we waited in our room for someone to come and move us all on to the next piece of nonsense, we were made aware that we had new neighbours, when the sounds of a drill instructor carrying out his calling drifted in through our window, and 'Horace', who was gazing out in idleness from this vantage point, aptly summed up the situation for the rest of us by remarking quietly, "Yo Ho! Crumpet."

We quickly joined him to see what was going on. It became apparent that we had new neighbours in the previously empty barrack blocks far across the other side of the parade ground, and, what is more, they were gloriously female. These fine examples of young womanhood were being commanded by a young and aggressively-voiced male corporal. Women never quite look their best in uniform when they are being drilled, or perhaps I should try and re-phrase that. With their top half naturally projecting forward and their rear being thrust backwards, they always tend to remind one of geese on the move to my mind. However despite all these aesthetic deficiencies, our youthful interest was immediately aroused and, young gentlemen electricians under training that we were, we began to offer verbal encouragement and the like from our lofty vantage point.

Before very long a red-faced and irate drill instructor halted his ladies' squad outside our barrack block, marched up the stairs to the top floor, entered our room and invited us none too politely to join his group in a spot of joint drill. This was of course exactly what we had been seeking, only now, after all our noise, we could be quite sure of being observed by the rest of our entry from their windows, and, what is more, could rely on them to play their part in the unfolding drama. It began to get out of hand at once, as you might guess. Women find it difficult enough to adjust to drill and marching as men do, and we were already experts. Now, with apprentices being marinated rather than blended into their ranks, it soon degenerated into something far worse, as the poor girls were now additionally required to do their best to avoid our not so furtive physical advances. There were at first some lewd suggestive whisperings and general leering, followed by light harmless heel tripping, which soon progressed to quick pinching and nipping of young ladies' bottoms when the drill person's eyes were elsewhere, and not a few squeals and little screams emerged from within the ranks, all to the generous and appreciative applause of the rest of our fellow Fifty-Fifth apprentices from the now crowded windows of our barrack block, and a few well seasoned shouts of advice to boot. The drill instructor still hadn't learned when enough was enough, and that he most certainly was not on a winner, and he again halted the squad while he went inside to recruit an even larger following from the many spectators at the windows. They would have been delighted to accommodate him, I can assure you. He was unfortunately waylaid by our disciplinary sergeant on the way up the stairs and advised that this was perhaps not an entirely wise decision. He re-emerged red-faced, dismissed us and marched his young ladies away to the safety of the far side of the parade ground. In one long row of twenty two, the Fifty-Fifth electricians took their curtain call bows to a rapturous reception from their fellows in the dress circle and private boxes above.

As you can see, by this point in time our entry identity had rather clearly emerged, largely based on the good offices of our seniors the Fiftieth entry and substantially reinforced, I might add, by some of the stronger of the emerging personalities of whom I have already spoken.

The instant multiple push-ups on the physical training sessions, the doubling round the perimeter of the parade ground on drill sessions, and the extra fatigue duties had now been replaced by the much more formal punishment of being put on a charge and awarded multiple days of defaulters by your commanding officer which was entered on your service record. This did not deter in the slightest the more flamboyant of the characters, such as Ernie Baldwin, 'Ianto' Bramwell, Tom Pickworth and others of all trades, who took all this in their youthful stride and showed the way for the more cautious and timid. They simply could not be disciplined by the threat of punishment, and would happily exchange three days' jankers at any time for just one act of riotous farce which left some member of the 'they' red faced and exposed to the laughing ridicule of the mass of us. There were some shallow imitators of course, but the real characters achieved a degree of affection and respect, as well as notoriety, which still lasts to this very day. The stories of which of them had done precisely what and to whom on any given day spread through the other barrack rooms like wildfire in the evenings, and gave much comfort. In this, I believe, lie at least some of the roots of that uniquely apprentice style of humour, completely non-understandable by others, which can somehow find someone else's extreme bad fortune excruciatingly funny, and to happily accept it without resentment when it is yourself who is the unfortunate object of such hilarity.

Within the context of one's entry identity was the closer, warmer and more familiar identity of one's fellow trade, room-mates. This was inevitable really with so small a group living, suffering and working so closely together. As you would believe, there was quite a mixture of personalities among the twenty two Fifty-Fifth electricians. Looking back with a more mature eye, I would say that affable Alan Jensen, was certainly the most vulnerable to having his leg pulled in those days. 'Slash' Gwilliam, he of the controllable colon and football skills, discovered this quite early on, when, with a twinkle in his eye, he speculated that Alan's Christian initial A perhaps really stood for Asser. Alan was far too quick to rise to the bait and replied indignantly that he was most certainly not Jewish. After that everyone tended to join in when the mood took them. Pat Cropley, who was to become a very good friend after we had graduated, was certainly the most violent at that time, with a penchant for lightly, and sometimes not so lightly as the mood took him, head-butting you between the eyes in what is known as the Glasgow kiss, and giggling as you went into a rapid recoil. Johnny Hopper was assuredly the most shy and intensely private, and even then it was completely inconceivable that he would ever marry, for example. 'Dai' Evans was undeniably the most light-hearted and boisterous, but in hindsight he was probably the one most likely to be eventually hurt by life. There were plenty of individual friendships, for example Ken Smith and Alan Jensen were always close, and 'Horace', 'George' Hanchett and I still seemed to find each other's company agreeable from the first induction test days. There were a corresponding number of dislikings too, and from time to time, particularly in the first year, these could erupt into fist fights. 'Horace' and 'Swing' Swoffer's small contretemps was just one example, and Pat Cropley's eruption against tiny 'Chiv' Chivers and threatened repetition another, but more and more our common situation tended to bring us closer together rather than emphasise our differences. In this way, 'Pash' Page and I never actually got around to coming to blows, but we did manage to spend three years actively disliking each other, and on reflection I can hardly bring myself to blame him.

SOCIAL CALLS AND VISITING FRIENDS

And when we're in the mush
There'll be no more shuftee...

A Fiftieth entry meeting in the 'tank' led by 'Ianto' one night decided to relieve the monotony by raiding the whole of both number one and two wings, a somewhat ambitious undertaking even for the Fiftieth. Robin Berry, still out of breath after running all the way from the 'tank', brought us the news. The Fifty-Fifth were to be allowed to take part but were only permitted to take on their juniors of the Fifty-Sixth entry, but the memory of that earlier remark of 'shower' by the Fifty-Fourth still rankled and virtually guaranteed that their rooms in number two wing would be receiving most attentive visitors. En masse the four hundred of us slipped silently across the intervening School of Cookery area and started on number two wing first. The first to encounter us was a lone piper practising at the back of number two wing. His tune ended in an agonising wail as the Fiftieth advance guard got him. The element of surprise aided us, and, after reducing their rooms to rubbish no matter which entry they were, we soon crossed the parade ground square which separates one and two wing ready to try out the second half of the operation. The chaos of tipped beds and wrecked kit lay behind us but, as one might expect, we had more trouble with the by then fully alerted number one wing. Finally we were obliged to discontinue the assault and to retire back across the parade ground. By this time the whole camp had been alerted, and we discovered that across our path lay a linked arm line of all kinds of 'they': drill instructors, 'boggies', disciplinary and Regiment N.C.O.s, and R.A.F. policemen, all our natural enemies of course. Behind them, waiting, was the whole of number two wing seeking a second crack at us, and behind us were our pursuers from number one wing. The situation appeared to be rather critical.

Then, like the miracle at the parting of the Red Sea, all of us apprentices of all wings and all entries suddenly remembered exactly who our real joint enemy was, and, what is more, just how thin and fragile their line was. Both sides simultaneously attacked the centre in a classic Cannae double envelopment, and many old scores were settled that night, I can assure you. In the midst of this gigantic punch-up, when it was discovered that the police had some of our number in the guardroom under close arrest, 'Ianto' personally led a group which stormed the place and released them. It was a most satisfying evening.

On the following day's morning parade, our Wing C.O. asked for those who had taken part to take one pace forward. Without a second's hesitation two whole entries, over four hundred apprentices, then obliged him. The punishment was to be confined to camp for a month and to pay for all the damage caused. You tended to get off lighter in the mass than you did as an individual, you see.

The Fiftieth didn't stop here. They used the occasion to organise entertainment for every single night of that month. There were football matches North v South, sing-ins, and all manner of sporting events so that it became a kind of mini-Olympics. The masterpiece was the wing dance. In the final year, an entry was permitted to attend an apprentices' dance, which was always held in the number one wing gymnasium. The girls were brought in by buses from the local villages. The Fiftieth 'borrowed' the keys of the number three wing gymnasium, and decorated it in secret. They then intercepted the buses on their way to number one wing and brought the girls to three wing and held their own dance. Of course it resulted in two more weeks' punishment but you had to admire the style. Of course the Fiftieth's instructions that we should only attack our juniors the Fifty-Sixth had proven quite impossible to comply with, particularly in the case of our friends the Fifty-Fourth, and, because of this apparent inability or unwillingness to recognise just who were their seniors, the Fifty-Fifth entry gained a reputation that night of the raid for being exceedingly cheeky bloody rooks who should be put firmly in their place at the first opportunity. Just as soon as their protectors the Fiftieth had graduated of course, that was only simple prudence. By the time that eventually happened, however, we had little to fear from anyone any more.

THE SEASON OF GOODWILL

He called for a light in the middle of the night

Our group identity as electricians within the Fifty-Fifth entry developed strongly in the next few months, in a way that I do not believe happened in quite the same way in the much larger groups of airframe and engine fitters. Being divided into many more separate classes, and housed in separate rooms gave them few opportunities to identify themselves so closely by trade as we electricians did, nor to develop a common character. As Christmas time approached, our C.O., the dog owner, announced that his good lady was going to bake a cake and it would be presented to the room with the best Christmas decorations. The electricians of my entry had developed into something that might be described as rather a bolshie group by the end of their first year of service, with no great respect or love for our masters. Our interest in this competition was therefore somewhat lukewarm and muted, to put it mildly. At a rather late hour, it was decided that we would perhaps have to make some minimum effort in order to avoid punishment for not even trying. It reluctantly began with the idea that spots of white blanco on the windows could be made to simulate snow flakes in the absence of any locatable cotton wool. No one had as yet found any reason to attempt to gain access to the sick quarters via the central heating ducts, you see. Each one of us became responsible for his own window, but the dip and dab soon palled and became more a dip and flick. The resulting streaky mess could then only be recovered by painting the whole window white, and this meant that the lights had to be left on all day to penetrate the gloom. We next considered coloured streamers, but they were well beyond our financial resources as well as our level of interest, so an attempt was made to colour stolen toilet rolls with inks borrowed from the C.O.'s office, and use these as a substitute. At best it could not be described as a roaring success. The only colours available had been blue, black and red, which can hardly be considered as gay or festive. Brown was of course also an option, if one was prepared to collect used toilet rolls that is, but was reluctantly abandoned in the interests of general hygiene. Coloured toilet rolls wound round lamp flexes however, no matter what the colour, unfortunately still tend to look like toilet rolls. The matter of a tree of course presented no problem at all, with the adjacent richly wooded slopes of the Chilterns just outside our barrack blocks. Although it had been expressly forbidden, of course, under pain of death and mutilation, six of the strongest of us departed into the woods which grew up behind our barracks, selected what appeared to be a suitably foliaged type and hacked it down. When it was dragged up the three flights of stairs and into our room, it was discovered to be somewhat too tall for our ceiling, and while attempts were being made to reduce its length with a fire axe, those holding it secure inadvertently pushed when they should have pulled, and the tree top broke through the plaster of the ceiling. Little else was then possible, other than to thrust the tree upright, and there it stood at the end of our room, majestically rising through the hole and up into the darkness of the loft above. At that point we gave up on the whole wretched enterprise in which we had never had any desire to participate anyway, and settled back for what was most obviously going to be our inevitable fate.

Accompanied by his good lady carrying the cake, the great man visited each room in turn before making his decision. As the top floor, we were fortunately the last to receive his visit, otherwise I am quite certain that the cake would have been withdrawn from offer and taken home for consumption, or perhaps even given to the dog. In the wise sudden absence of our Fiftieth entry apprentice N.C.O.'s, who were astute enough to recognise impending trouble when they saw it and had no great wish to become personally involved, one of our number whose bed was located near the door called the room to attention as our leader gently ushered his lady through our humble portals, as an officer and a gentleman should, and he realised his mistake at once as he peered around in the vast gloom in sheer disbelief. We had been considerate enough to turn the lights out you see, and in any case it was strictly forbidden to have them switched on in the daytime.

"Turn the lights on at once!" he ordered, and then of course the terrible truth was revealed. It was, however, now far too late to extricate either of them from what was all too obviously, very very wrong.

"What on earth is that mess on the windows?" he enquired, pointing his finger.

"That is snow, sir," the apprentice replied.

"And what are these?" our leader enquired, this time ever so slightly more noticeably irritable, now indicating with his pointed finger the toilet rolls which had now become the centre of his attention.

"Streamers, sir," was the reply.

"I suppose you are going to tell me that this is a Christmas tree?" he said, advancing menacingly down the room. "And I will certainly want an explanation as to where it came from, but what the hell is that hole in the ceiling for, apprentice?" he shouted, having now plainly lost his temper, even in front of his equally shell-shocked but still woodenly smiling good lady who was stood in the doorway clutching the intended prize for our humble efforts.

"For Father Christmas to come through, sir," replied our ever-resourceful apprentice.

It had not escaped our leader's attention that one of our number was 'Horace', and having harboured certain suspicions for some time that his dog's sometimes bizarre behaviour was somehow linked to the near-presence of that individual, all this did nothing to reassure him that this was not some kind of deliberate evil plot directed against him personally. We did not win the cake, just as we had realised and accepted from the very beginning, and punishment was inevitable, including yet another bill for damages, but from then on our C.O. watched us with a very wary eye indeed, as if we the electricians were some kind of terrorist faction planted within his command.

Another strange custom came into being at about that time. After lights out one night, someone told some amusing story one night which appealed to 'Dai' Evans and he began to laugh in his best hearty fashion. This struck some sympathetic chord in 'Slash' Gwilliam and he began to laugh more and more until the tears rolled down his cheeks. It spread like wildfire from one to another until everyone in the room was convulsed, but had no real idea any more what they were laughing about. Even the intervention of the orderly sergeant on his evening rounds could not stop it for some time, and ever after over the years it was quite likely to break out again, and usually for the most flimsy of reasons. We were all literally on the brink of some mass hysterical insanity it seemed, but it was funny.

It is traditional in the R.A.F. that the officers and N.C.O.s serve the airmen, in our case apprentices, their Christmas dinner. In our case this was adjusted to be before we all departed for our seasonal leave. It was and is, a time of bonhomie, much enforced merry-making, and general pretence about what all-round good chaps, and beneath any badges of rank, deep friends we all are, for a few hours at least.

On this occasion we were treated to an unsuspected extra ritual, when at the height of the festivities, a Fiftieth entry representative suddenly got up, and insisted on presenting seasonal gifts to prominent members of the officers and the N.C.O.s. They were hardly in a position to refuse after all, although several noticeably paled. The recipients had to open the gifts after being called forward, in front of everyone of course. The gifts had been cleverly chosen to be symbols of great significance to all the apprentices, but completely lacking in symbolism to the recipients. The Wing Commander, who was known by us all for his slow-moving cold repulsiveness as 'The Slug' opened his parcel to find a cabbage leaf, and was heard to enquire of his attending adjutant, "I wonder what the significance of this might be?" Had he examined the underside of the leaf a little closer he might have had a better clue. The disciplinary Flight Sergeant received a set of false teeth in blank non-comprehension. Had he had the wit to examine the dentures further, he might have noticed the piece of chewing gum which was attached. The message was that he needed something to keep his mouth shut. The Fiftieth were simply never ones to neglect an opportunity like that.

It was about the end of our first year that we were gathered together in the 'tank' to hear of where those who had been earlier promoted to N.C.O. apprentice in our entry were to be moved when the Fiftieth departed and the Fifty-Eighth arrived. As early as the first summer there had been three promotions to leading apprentice but they seemed to have been based on recommendations mainly from workshops for outstanding craft skills. They were soon followed by more including 'Ig', Robin, 'Pussy', 'Curly', Ken and Eddie LeGrove from the electricians and in the final year Alan Jensen, and although I did not know it at the time, these latest were based on recommendations from two out of three from the list submitted separately by the Wing, Workshops and Schools. There were not too many big surprises in the later batch however. Our friend the dog handler could at last abandon his hairy charge, and the human windmill could begin to shout at some one else, rather than be shouted at, so there had clearly been some Wing influence. From the six promotions amongst the electricians, with the exception of perhaps one, I had not expected any of them, or the rest of us, to be promoted either. The top three electricians at Workshops were not represented, nor the best at Schools. If you had chanced to have clean buttons on the day that they were particularly looking at buttons, or you were the one chosen at random to march a group from A to B as the Wing Commander passed by, and he had failed to notice the central finger of the right hand raised in your salute, then you might well have got a good Wing rating. Those with hobbies in some acceptable club stood excellent chances, whilst all those who had joined the pipes and drums could safely abandon any dreams of promotion for all time. It was very rare for a band member to be seen wearing any N.C.O. apprentice chevrons. They were most clearly considered to be 'skates' on that premise alone. It was quite disappointing to notice that in fact the 'they' did not really know us very well at all. Despite my original enthusiasm for the service, I was already quite sure that, whatever the criteria used, I would never have been selected, nor did I expect to be in the future. The privileges for the promoted were to enjoy a thirty-six hour pass every month, to have a priority for promotion to corporal after apprentice time, to have their own room, and the right to abuse others in turn, and some did just that. I came across one of these when I was the newly-promoted corporal temporarily in charge of his room when we were in our improvership year at St Athan. His anxiety showed all too clearly what was passing through his mind, but nothing further needed to be said between us other than my private quiet comment that now perhaps it was my turn.

GROWING PAINS

And the woodpecker said God bless my soul
Take it out. Take it out. Take it out. Remove it

A whole year of service had now elapsed and about the beginning of the second year was what might be described as crunch time for most of us. Every single one of us, believe it or not, had started out starry-eyed and determined to do well and to prepare ourselves to take our places as members of what we thought to be a technical elite, but most had been obliged to come to terms with the day-to-day realities of our true situation. It was most certainly not going to get any easier for the next two years at least, and, who knows, maybe not even then. The more distant future was certainly none too inviting either if you cared to consider it from a probable career viewpoint. There were lots of engine and airframe fitters on long term engagements already working out there on the squadrons in the R.A.F. with only one flight sergeant and maybe one other sergeant in charge, so there were likely to be very few promotion opportunities for the majority, unless you were lucky and someone died unexpectedly early of course. Most of the extremely competent and skilful practitioners of these trades found few promotion opportunities in their later service. The ancillary trades with a flight sergeant per camp and a sergeant per section would probably, and did, tend to do a little better, but only by reason of their relatively smaller size, and indeed in later service you invariably knew the successor who would be taking over from you. When I finally departed the R.A.F. for example, I knew that 'Uke' Lailley, ex-Fifty-Second entry, would be taking my place as i/c ground electrical section, and even I felt some pity for the poor Air Force then, because I knew that my young engineering officer was enjoying high hopes about the arrival of someone better and more compliant with his wishes than I had proved to be, but from my advantageous position of more personal knowledge I also knew full well that 'Uke' was unfortunately just the same if not in fact worse than myself.

At this time in our apprentices' life, however, no one was actively seeking to be a problem boy yet, but there was a great deal of disillusionment showing if you cared to look very far. It was becoming increasingly necessary for many of us to find some way to relieve the constant oppression and petty injustice. The earliest dissidents seemed to come from those with a well-developed sense of self rather than a group identity and were by no means from the ranks of the less intelligent or criminally inclined either. 'Scouse' Fergusen for example, was one of the three earliest promotions because of his high technical skill level as an instrument-maker, which was no small thing in what was one of the more highly demanding technical trades, but he was not long in abandoning what he obviously considered a rather dubious privilege. Another was my good friend 'George' Hanchett, who had been serving a civilian apprenticeship at B.O.A.C. as an electrician and was technically outstanding. The Air Force was very quick to clamp down very hard on anything which even faintly smacked of potential insubordination. Obeying was and remained, very much the in thing, and of course this did not sit too well with those of us who preferred to at least think, if not act, for themselves, and inevitably this led to head-on clashes from such a wide divergence of viewpoints.

Any form of extrovert behaviour was considered by the service to be a sign of either active subversion or someone attempting to 'work his ticket', meaning to get himself discharged. In either case it seemed that it was to be ruthlessly eliminated. There was plenty of extrovert behaviour going on however, but mostly just for the sheer joy of it. One of the Fifty-Second entry electricians, Dave Doubleday, specialised in terrorising the passengers on tube trains when going on leave. He would set fire to the opened newspapers of bowler-hatted city gents or, as an alternative, would growl like a dog and then suddenly seize the lapel of any dark-suited gentleman in his teeth and shake it violently. He was never vicious you understand, and always smiled happily as he went through his performance. When going on leave on one occasion I once arrived on a tube station platform just in time to see the crowded rush hour train already departing. Except, that is, for one compartment where, stood alone in majestic solitude was our good old smiling friend Dave, with the other passengers huddled together like sheep in a pen at the farthest end possible. He had obviously started his act rather early that particular day. It was certainly far better entertainment than the present-day rather poor musicians and magicians who frequent the Underground. One simply had to adjust to Dave and make some allowances for his ways of expressing himself. When I was the corporal in his room at St Athan during the improvership year, he would invariably wake at reveille, stretch and then begin his day with a rising booming crescendo which made quite sure that absolutely no one overslept in our room.

"Up! Up! Up! In the morning early," was his daily cry. This did tend to alarm some of the less experienced non-apprentices I will admit, but not myself or others who had known him somewhat longer. 'Dusty' Miller from his own entry summed it up so well once when he remarked, "He was a bloody nutter."

The sudden impulse at the most unlikely times to do something, anything, came over most of us at one time or another. I once saw 'Horace' suddenly pick up his blanco brush and paint the toe caps of 'Slash' Gwilliam's boots white only minutes before an inspection on just such an impulse and the expression on his face showed all too clearly that he didn't really know himself why he had done it.

STARTING TO DO MY SHARE

*Now pack your bags, and luggage too
And go live on the...*

At workshops we had progressed through some basic electrical classes such as cable jointing, field telephones, batteries, direct current and alternating current machines, and on to elementary aircraft wiring and ignition systems. Later as the months turned into years the cyclic repeated subjects became deeper and more relevant in our future role, and began to include such joys as aircraft power systems, automobile and marine craft wiring, bomb gear, airfield lighting systems and flight simulators. There were many roles that the R.A.F. electrician could be called upon to fill on a working station in those days. Later the trade was split up into air or ground electricians to overcome this divergency of equipment problems, and we were in fact the last entry to be required to handle both options. The Fifty-Sixth, after having being trained for two and a half years on both options, had to make their own selection of which one to retain, bare months before their trade test and if successful automatically became Junior Technicians in that trade. The length of these repeated-through-the-years mini-courses was usually about two or three weeks, which was usually just about sufficient to allow one to begin to get bored, and to seek a spot of relieving mischief.

Sometimes the instructors were civilians like our friend Mr Tripp and sometimes R.A.F. servicemen instructors. One could not assume that the civilians would be any more lax or easy-going than the servicemen, and, if discovered in some small piece of foolery, punishment invariably came hard and fast from either. However, sometimes a spot of decency managed to prevail even in this Dickensian outpost, depending on the character of the instructor.

One of our early civilian instructors teaching D.C. generators was a thoroughly decent fellow with a deep interest in cultural matters, which were unfortunately entirely out of keeping with either his philistine pupils or his job. Although not Jewish himself – in point of fact he was a devout Catholic – he could speak and write Hebrew for example, and in fact he kept his class reports on us all in this language, so that for once we did not discover his rating of our miserable efforts, even when we had succeeded in picking the lock of his desk as we did with all the other instructors. Even so we managed to provoke even this calm generous-spirited man into one explosive outburst, as was our intent of course, when one of our number, in a desire no doubt to express our own cultural depth, had chalked 'coitus interruptus' on the inside brim of his trilby hat. This had surprisingly remained undetected until it had apparently caused a small stir amongst the ladies at his bus stop when he had left work one evening. The following morning in class the smoke was still noticeably emerging from his Catholic ears after the unfortunate affair of the previous evening, and as he was writing on the blackboard he must have heard our giggles and whispered repetition of the Latin phrase behind him. He turned around angrily and told 'Horace' to leave his classroom at once. Poor 'Horace' was not in fact a particularly suitable choice for dismissal on this occasion, for once being completely innocent, but out he had to go despite his loud protests. This presented him with some serious difficulties as to precisely where to go and hide until the next break time when he just might be readmitted to the class, because anyone discovered just hanging about anywhere at all, including the general workshops area, was in for some real punishment, and the toilets located just outside the workshops were freezing cold and equally, if not more, inhospitable than the bays themselves. Ever-resourceful 'Horace' was not easily defeated however and simply removed his overalls, put on his hat, took his notebook from his side pack, tucked it under his arm and set off for a long tour of all the workshops of all trades at a brisk march, as if seeking someone in authority with a report.

On a later second visit to our kindly instructor we were now being taught the mysteries of electric motors, and on the point of being given some practical experience exercises after many hours of tutorials. However, such had been the impact of the previous hat chalking incident that he could not quite bring himself to trust us entirely, at least in the mass, and accordingly he selected those whom he obviously considered to be two slightly more responsible members of our class, 'Ig' Noble and Robin Berry, to carry out the practical aspects for all of us, probably on the simple premise that they were wearing apprentice N.C.O. stripes and were therefore comparatively trustworthy. They were each separately entrusted with the repair of one electric hand drill whilst the rest of us sat and awaited their completion. Finally, when all was apparently ready to be demonstrated, 'Ig' obliged at the instructor's behest by pressing the trigger switch. Absolutely nothing happened at all and many frantic pressings only produced low giggles from the rest of us. 'Ig' then looked anxiously at Robin who pressed his switch, only too willing to try and divert attention from his unfortunate comrade. There was a bright flash and a simultaneous bang, and a plume of smoke rose from the drill. All the rest of us simply fell about laughing. The poor instructor never let any of us loose on any piece of equipment whatsoever for a considerable period of time after that. There was a small sign of our true regard for him much later in our final year, when we discovered by chance that one of our number had stolen an electric motor from our distinguished instructor's classroom for which he would be eventually required to pay. That night after lights out, the culprit was forcibly persuaded to return it before anyone was the wiser.

Whilst we, the gentlemen of the 'A' class, were under instruction in one subject, our 'B' class colleagues would be located elsewhere on quite another. On one occasion they had been learning the deep mysteries of portable electrical power systems, which were diesel engine-driven versions of the systems in general use in civilian power stations, and could be used to supply power and lighting to whole camps if necessary. One of my classmates was eventually required to set up such a tented camp completely from scratch during his Middle East overseas posting. Can you possibly imagine that degree of responsibility being given to a twenty-year-old today – to set up on his own initiative the total electricity supply for what amounted to a small town? Incidentally he went on to achieve a double degree university standard, and enjoyed a distinguished career in the electrical supply industry after leaving the service. Perhaps we were being prepared for much more than we could possibly imagine at the time, and it is also a sharp reminder of what the Air Force has so recently chosen to abandon.

Part of our 'B' classmates and our own early training was the means by which two or more such separate generating systems could be brought on line together. Without going into excessive technical detail, there are a number of factors which must be adjusted to rather precise equivalency in the systems separately before the switch can be thrown to connect them as one common source of power, and there are a number of alternative aid methods to ensure that this is successful. Failure to do so can be catastrophic in the extreme, I must add, and has razed power houses to the ground on occasion. As their instructor at the time put it, "It must be spot on."

Our classmates, after practising these methods under the supervisory eye for a while, began to be at least mildly curious concerning just how on had to be spot on, and then what happened precisely if you selected to get it spot off. This they began to experiment with when their instructor had begun to have sufficient faith and trust in them to depart for an early smoke and a cup of tea, leaving them unsupervised.

We of the 'A' class at our remote location detected this when the lights in the workshop bays began to sometimes noticeably dim at unexpected times, usually ten minutes before a break time. Then came the day that the lights dimmed to almost total extinction and remained so, and a deep rumbling like an earthquake shook the very foundations of the concrete floored bays, later reported as reaching as far as the engine fitters' bays on one side and the armourers' and instrument makers' on the other. Our own instructor stopped his discourse and looked up in alarm.

"What on earth was that?" he exclaimed.

"Oh don't worry, sir," someone answered lightly from the back benches. "The B' class are on practical this week."

When our own turn came to study this subject, the same instructor had by now realised that somehow he would have to try to arouse and sustain our interest, or we also might begin to be technically curious. He did so by a method absolutely guaranteed to catch our undivided attention. First he pointed to a very recently painted notice in bright red hung in a prominent position on the classroom wall which read, 'Care and Caution are Essential, if you work on High Potential!' Then he went on to recount his own experiences in some far foreign field, where it seemed that a careless young airman had taken a member of the women's arm of the service behind the switchboard of just such a vehicle as we were now studying mark you, for a certain obvious purpose and, disregarding the notice's warning had suffered a literally shocking fate. Getting at once to the technical heart of the matter came the question from his audience, "How do you know what he took her there for sir?"

"Because of where the burns were," he replied.

The aircraft wiring systems entailed some practical work which, without the benefit of real aircraft, required one to mount components, and wire up particular types of circuit on the large vertical partition wooden boards, whilst the reverse side was being similarly used by someone else. I wish that it had proved to be so easy later within the limited confines of actual aircraft. Then exactly how to manoeuvre your body in order to get your hands on something or other proved to be at least half of the job in hand in my experience. The wooden breadboard approach, however, presented limitless opportunities for the more mischievous to drill a small hole through the board, and to make some unfortunate changes to the masterpiece of the one on the other side. I overheard Robin Berry exclaiming with surprise one day that he actually was able to feel the twenty-four volts of the battery supply. A glance revealed a happily smiling 'Horace' on the reverse side of the board winding happily away on a 250 volt generating mega-insulation tester, the leads of which disappeared through just such a small hole in the wooden board. I myself found a long thin nail and carefully hammered it through the board and into the back of some of 'Slash' Gwilliam's wiring on the other side. It took both his and the instructors' combined skills to find out exactly why the fuses blew every time the supply was switched on.

As a small relief from all this hard work, one day I announced my intention to beat retreat with the 'Grim's' pipe band. From my small pack and various lengths of thick black starter cable joined by the braided coloured core cables from some light wiring, I manufactured my set of pipes, and began my march up and down the rows of circuit boards being attended to by my classmates. Flight Sergeant 'Dadda' Brett, the electrical workshops supervisor, happened to be passing and noticed the tips of my pipes passing above the tops of the partition boards as I held my small parade. His curiosity thoroughly aroused, he rapidly moved to the end of the board and peered round, fortunately just as I had turned the opposite end, out of his sight. I was of course then given good warning of his approach by my comrades, and when he followed me round the corner he found me carefully mounting another piece of wiring. Still puzzled, he departed shaking his head.

I spent that first Christmas leave at home and then went with Paddy Swoffer to Guernsey in the Channel Islands for New Year. Paddy had grown up there and his father had been the airport manager until his death, but the family no longer lived on the island.

Paddy still had many friends there, however, including the two sons of the Bailiff. It was a thoroughly enjoyable holiday and we decided to extend it a trifle by missing the daily boat to the mainland. Paddy presented himself at the harbour R.T.O.'s office all out of breath and suitably anxious, after watching the boat sail, explained his unfortunate lapse of timing to the sergeant in charge and was given a chit. I didn't feel that this was quite honest, so I did not bother to accompany Paddy to the office. When we returned a day late, we were of course immediately put on a charge for being absent without leave, and eventually brought before our C.O., the very same who had most recently visited our humble abode with the special Father Christmas entry point. Paddy then produced his chit. The charge against him was dismissed, and he was marched out. I had no chit of course, but the C.O., assuming that my situation was the same as Paddy's, began to reproach and then abuse me for being so foolish as to assume that Paddy's chit also covered me. I did not hesitate to put him right on the precise truth of the matter, and duly received my first three days' defaulter's punishment, known in the Air Force as jankers.

This involved having to get up early in the morning and reporting at the guardroom in full webbing equipment, for inspection by the duty R.A.F. policeman. He could be thoroughly relied upon to be in a foul mood at that hour of the morning after being on duty all night, and just looking for someone to sink his fangs into. In the evening after work, there was an early extra parade for flag lowering, following which you were given an hour's fatigues work. After that, for the remainder of the evening, until lights out, you paraded at hourly intervals in a different dress for yet more inspections. These were usually so exacting that there was more than a strong likelihood of being put on a charge yet again. The main difficulty in being on jankers in fact, was how to get off them. You were also required to wear a white cloth armband on the left arm above the elbow, much in the fashion and for similar reasons that lepers were once required to wear bells. This made you even more of a marked man, and very likely to be called out for punishment when the true culprit in some criminal activity could not be identified. The call of, "We cannot be sure who precisely is responsible. Valance, Bristow, Hanchett, one pace forward march," became an all-too-familiar event. The steepness of the hill from the guardroom up to Maitland barracks became even more familiar to me in all seasons, and in all weathers. Every day I marched down it to work in the morning and afternoon, and back again at lunch time and in the evening. Every G.S.T. afternoon, I ran up and down it in P.T. kit, or clad as an infantryman. Every leave time, I marched down it to the railway station. Now I trod an even deeper path, with my individual visits to the guardroom in full webbing equipment.

Apart from this, life went on as usual. At schools one warm spring day in the instrument-makers' class, the teacher, a young Pilot Officer, was writing on the blackboard, and did not notice that one of his pupils was idly watching the passing airmen from the open classroom window. He did not hear the greeting, "Hello Sambo," which was directed at the passing Jamaican airman either but he did notice when the door burst open and he was obliged to intervene and try to separate a by now very uptight and shouting coloured airman and an innocent-faced boy. This kind of 'bar rattling' helped to relieve the monotony you see, although the risk of swift retribution was very high. The instrument-maker in question, Roy Ellis, lost his recently awarded leading apprentice stripe as a result of his brief moment of racist 'pleasure'.

The instrument 'bashers' were in fact going through rather a general nadir in their fortunes at the time. 'Scouse' Skelly, even smaller than our Robin but a great little boxer nevertheless, and with the typical impish sense of humour not entirely unknown in his city of origin, had been lifted and hung up on the clothes hooks outside his schools classroom by his classmates one day. Perhaps it was something that he had said or maybe it just seemed to be a good idea at the time. Unperturbed as ever he had taken out a comic from his pocket and was reading it in this rather unusual position when the education officer had arrived.

"What on earth do you think that you are doing there, Skelly?" he had asked.

"Oh don't pay any attention to me, sir," 'Scouse' had replied. "I'm just hanging about."

He got three days' defaulters in which to hang about further that same afternoon.

The 'boggies' of the cookhouse staff had for some time been increasingly inclined to a certain degree of laziness and contemptuous hostility to us, no doubt brought on by the frequent allocation of our young bodies to the more menial of what would normally have been their tasks. We decided that it was perhaps high time to right the matter. There was in theory an evening meal called supper, where a mug of cocoa and maybe a sandwich or so should be distributed for those who wanted it. Few now did since, over time in practice this had deteriorated to a slice of bare bread and the remainder of the tea, now cold from the earlier after work meal. Accordingly all two hundred of us turned up one supper time demanding our rights. A protracted argument with the duty cook and eventually his N.C.O. resulted in a promise that all would be better the following night when we returned. Those on jankers and fatigues confirmed early the following evening that indeed better things did await us at supper time, so consequently none of us turned up at all, but the following night we all did, and so a cycle started for a while. It was remarkable how short a time elapsed before we managed to catch their undivided attention.

From all the wing personnel's selections of potential leaders, there was not one of them with a real parade ground voice that could be heard by two hundred apprentices on the march, and the entry was obliged to elect one of its own, an airframe fitter called 'Mo' Moran. He was a stocky little fellow whom you would never normally notice but any football club would have been proud to have him on their terraces with his bull horn voice. If the entry was on the march, he would move to the front and absolutely everyone heard the orders. Although they didn't like it, since he was not one of their choosing, the 'they' began to have to accept it and 'Mo' would be reluctantly called out to begin the drilling.

The wing personnel had steadily worked at eliminating any individual acts of defiance and avoidance of their Saturday morning, or indeed any, of their routines. The days of 'Horace' and his reporting sick, and 'Swing' and his loft-hiding had long gone. Roll calls and multiple checks had made it virtually impossible for anyone to be absent from the drilling and P.T. However the electricians viewed this as something of a challenge, and decided that, if they acted as a group rather than as individuals, then it was highly likely that the absence of all twenty two of them, excluding the ones who had been promoted to N.C.O. apprentice and were now in charge of other rooms and entries, would be assumed to be part of some bigger plan, and be disregarded. One apprentice standing at the bus stop alone would certainly attract attention at once, but a large group marching there would be likely to pass without notice. So one Saturday morning, after the usual morning room inspection, the hypothesis was put to the test and the electricians formed up, one stepped forward and gave the necessary commands, and we all marched down to the bus stop, just in time to get on board and depart for the joys of Aylesbury. Of course we had nothing to do there nor the money with which to do it, but that was hardly the point. Our assumption unfortunately proved to be incorrect, and we were all put on a charge. Three of the hardened criminals, myself included, were given seven days, Terry Thornton, for no apparent reason, was let off scot-free to his deep chagrin, and the rest got three days. Not entirely consistent, you might think, considering that it was for the same offence. Of the promoted ones, diminutive Robin Berry, he of the tool box, was now in charge of a junior entry room as a 'snag', meaning leading apprentice, as were Ken Smith and 'Pussy' Funnel. Eddie Legrove and 'Ig' Noble were corporal apprentices in charge of whole blocks and 'Curly' Coppock was the sergeant apprentice for one of the squadrons in three wing, so the electricians had done rather well in the promotion stakes considering their rather small numbers.

GETTING AWAY FROM IT ALL

My parents they thought it unfit
The part I had chosen was rotten

My own resistance was beginning to rise from the mildly passive into the blatantly active. I had a similarly-minded friend, in 'George,' real name Frank, Hanchett and it was only a matter of time before both of us hit serious trouble. Disenchanted one day with the ongoing fiasco of schools, we and our instrument-maker friend 'Scouse' Fergusen, no longer a leading apprentice as a result of previous crimes, agreed by mutual consent at the morning break time to completely abandon the next class and seek solace in Wendover village, at the cafe. Just before lunch time, we walked back to the wing area via the back roads through the woods. As we got closer, we all noticed an airman on a bicycle who, when he saw us, dismounted and blew a whistle. From the surrounding barrack blocks then emerged a host of the 'they', armed with what appeared to be pace sticks and other lengths of wood. They then set off in our direction at a fine pace. Although it was patently obvious that they were well aware exactly who we were and that therefore the game was most definitely up, our automatic reaction was to run and to split up in three different directions so that it was at least somewhat harder to catch all of us. In close pursuit behind me was my good friend the P.T. instructor 'Swede', blessed by the fleetness of foot quite usual in his chosen profession, with a large piece of wood in his hand which he brandished violently as he ran. I did not make any foolish assumptions that his intention was perhaps to use it to hail a passing taxi. We were sufficiently well acquainted by this time, after our many enjoyable joint jogging sessions together, to permit me to anticipate the likely sequence of events about to take place. He was after all basically a somewhat uncomplicated creature, with rather predictable behaviour patterns. A turn in the pathway allowed me to have a short breather behind a tree and as he approached I stepped out, and, as they say 'let him have it'. Although still rather small, comparatively speaking, I was already quite hard, and he released the piece of wood and sat down firmly on his bottom on the dusty pathway with a look of bewildered surprise on his face. I did not feel that I should presume on our acquaintanceship further by remaining to exchange further pleasantries about his general well-being and so on, so I continued on my way in a brisk airman-like manner.

Of course it wasn't long before I was successfully detained by others and I found myself reunited with my companions in number one wing guardroom in adjoining cells. Later in the day I was marched back to my number three wing home, and I was again in front of my C.O. This time I was given fourteen days' defaulters. The matter of the small physical tête à tête was not brought up on the charge sheet, probably to protect his pride, and to prevent any embarrassing enquiry from me as to the intended purpose of the piece of wood. After this I was marched in again and told that my commanding officer was sure that I would take this as a man. Three and sometimes five days' defaulters was not unusual, but twenty one consecutive days was, and he knew quite well that it would not be long before I was back in front of him for some further petty matter resulting from one of those nightly inspections.

I did not bother to attend the first six o'clock jankers call and register my presence, instead I departed for home the same night via the back window and the drainpipe. If I was going to be on what seemed very likely to be a long punishment, then it seemed better to have enjoyed a good rest first. By hitchhiking, I was home early the next morning on the back of an open potato lorry, and then I had another interview to face. The Air Force was always most careful to inform parents when to expect their sons home on leave, and for precisely how long. My father, therefore, was under no illusions at all concerning this surprise visit of the fruit of his loins. He was far more concerned it seemed, however, with what the neighbours might say when the police arrived to collect me, as they undoubtedly would in his opinion. If I had expected a sympathetic ear then I would have been disappointed. My difficulties, as always, were simply not explainable to him, so I didn't even bother trying. I had my pleasant weekend at home, noting well his increasing nervousness when the postman or paper delivery boy called at the front door, and the growing tendency to view the street from behind the safe cover of the living room curtains, and then went back to Halton by the same thumb-waving means as I had arrived, kindly sparing my parent the embarrassment of the police arriving in their siren-sounding blue-flashing-lamp cars to break down the door or shoot it out with me, and the neighbours giving interviews to the tabloid press which would be headlined 'I Knew Mad Dog Bristow'. The Wing Commander eventually wrote a charming letter to my father, informing him of my little unauthorised absence as if he didn't know, and eventual happy return, but the general tone was full of hope and good cheer, and even wished my father a belated Happy New Year. As a footnote it was suggested that he could if he so desired, purchase my discharge for a mere one hundred pounds. Since my father had never had a hundred pounds in one lump in his whole life, this was all rather academic, and so apparently neither of them took it or me very seriously. It must have been a great comfort to them both that, by the time all this eventually transpired, the object of their mutual affection and concern was firmly ensconced in the comforts of a cell in number one wing guardroom yet again.

It was already well after dark when I got back, and so I had no difficulty whatsoever in slipping into the camp unnoticed. There was something of a surprise awaiting me when I got to my barrack room, however. In my absence a reorganisation had taken place, and each trade had been split into two groups. One group remained where they were, and the other had moved to new barrack blocks to form a new squadron. The split seemed to have been made on the basis of desirables and undesirables rather than any other criteria and I, apparently being one of the 'hard lags', found that all my kit was now securely located elsewhere. My mates shared their blankets with me and I slept on the floor for that night. In the morning I soon located my new room.

Whilst I was recounting my adventure story to my fellow rejects, the new duty corporal arrived to call them outside to be marched to work. Since I knew with some small degree of familiarity what was supposed to happen next, I lagged behind somewhat and this immediately attracted his attention. When I began to explain, he silenced me with a shout, and told me to go and wait for his further attentions outside the squadron office downstairs. That proved to be for some time, while he carried out his marching detail orders, and the new disciplinary Flight Sergeant arrived at his office before the corporal had returned. He casually asked me what I wanted, and I replied that the corporal had told me to wait, not overly anxious to be too helpful in view of the response to my earlier explanations. He was by this time inside the office and already idly examining the papers in his in tray, while I was barely across the threshold. He didn't even look up as he indicated that I was to march to workshops alone, with his permission, and when I returned at lunch time the matter could be sorted out. Dutifully, I about-turned and began to leave. My hand was even on the door knob when he asked that so vital and obvious question, "What is your name by the way?"

My reply brought about the fastest reaction known to humans. I didn't even manage to get the door partly open before he was leaning on it with all his tubby little might and shouting, "Help! Help! I've got him!"

I was soon on my way, under escort, back to the familiar architecture of number one wing guardroom. There I was allowed to languish for some days, presumably to allow my mind to focus on the hideous nature of my crime.

My belt, tie and shoelaces were taken from me for obvious reasons, although I was not in low spirits by any means I must add, and I was locked in my cell all day, and only allowed out for exercise for one hour in the evening or the calls of nature. The guardroom corporal snoop was supposed to supervise this event, but in fact, being a lazy creature, he used to lock me in the inner exercise yard and go and put up his feet and practise reading his comic.

I had a companion in the next cell, an apprentice from a senior entry to mine. The stifling oppressive atmosphere of the training school was not exactly what he had been seeking in life. He was a boy with a rather exaggerated sense of adventure, and was being punished for his attempt to cross the English Channel in a stolen boat and join the French Foreign Legion. The coastguard had been obliged to rescue him when the yacht sank, which was how he had been apprehended. He was something of an eccentric, as you may have gathered. My cynical belief was that, even had he made it, he would only have found that he was joining the apprentices all over again.

It happened that whilst I was incarcerated there, Parents' Day came round. Mummies and Daddies in their finest were allowed on camp for the day to see how their little sonny was making out, and would walk proudly round the whole magnificent complex accompanied by their offspring, released for the day. The day began rather early for me, but I could hear the happy chatter through my cell window as the parents walked by the guardroom on their way to their meeting with their heir and offspring. My next cell neighbour saw this as a fine opportunity in which to indulge his sense of humour. He began with the plaintive wailing of one suffering and in pain:

"Oh, don't beat me. Please don't beat me any more corporal. I can't stand it." It went on, sometimes louder and sometimes falling to a bare whimper. Even I could detect the hush which came over the passers-by. He was making his point rather well, I thought. The surprised and concerned parents began to make enquiries when they arrived at the wing areas in search of sonny and it was not too long before my fellow prisoner, and I for good measure, were taken out of our cells and marched down to the coal yard for a spot of useful employment in shovelling, well out of contact with gentlefolk.

Sunday arrived and I was taken to church. This was not to ensure any salvation of my immortal soul, but because it was church parade day coming around yet again. My entry had been obliged to parade and be marched there as usual, in the ghastly monthly ritual. On this occasion, however, I came as a special guest, under escort by two R.A.F. policemen, and was made to sit between them at the back. No verbal contact was permitted with any of my fellow apprentices, but, as they filed past when they left, they still managed to drop cigarettes, which were known as 'spits', and matches into my hat unnoticed. In those days and on our pay, these were very much luxury items.

Soon after this, I was visited in my cell one day by the Padre. The same who had failed to give us the benefit of his fine tenor voice in the 'tank'. Whilst I was rather familiar with him, at least in appearance, he clearly did not know me at all, in fact he had to refer to his little red book, rather than his little black one with the gold cross on the cover, to even find out my name, but I was there on his list it seemed, as well as being in the big register in the sky of course. He assured me that he had known for some time that I was troubled, and that now he was going to help me. I will admit that I did enjoy for just a brief weak moment a fleeting vision of my punishment being reduced or an early release from captivity, but he then clasped his hands together, went down on his benders in the middle of the cell and prayed for me, before making his departure. I cannot say if there happened to be some failure in communication on this particular occasion, perhaps due to the unhealthy location, or if he was not as well connected in higher circles as he himself believed, but I do know that I stayed exactly where I was. In retrospect I think that they were jointly trying to decide what to do with me. I was seemingly oblivious to punishment, so I wasn't going to be broken that way, and yet my workshops training record was excellent and at schools I was at least middle of the road. There just had to be a satisfactorily simple explanation, and what could be better or more convenient than that I had probably fallen in with evil companions. Since the same thing seemed to apply equally well to my friend 'George' Hanchett, the only remaining question was if it was I who had been a bad influence on 'George', or was it that *he* was on *me*, but certainly someone was to blame and it couldn't possibly be them. The Padre's visit was just another probing ploy to find out what I did really want. I wasn't entirely sure myself, but perhaps some of my pre-Air Force experiences were being brought to the surface in this climate, and had at least something to do with my general take it head-on approach to discipline.

After a week, I was brought back to my new squadron and its brand new Commanding Officer and awarded another twenty eight days' jankers. This brought my total punishment so far to seven days' close arrest and fifty two days' defaulters. My new C.O. appeared to be a somewhat morbid introspective person, and, on this auspicious occasion of our first meeting, he did not express any opinion with regard to my ability to take this as a man or any other form of creature, as had his predecessor. It stood, however, as an entry record at the time and I was not entirely dissatisfied with that.

Others came to hold our new C.O. in less than endearing terms including the less criminal. 'Ig', though an N.C.O. apprentice, ran into great difficulty in getting a weekend pass to visit his girlfriend in hospital, and finally only succeeded with 'Steve's' assistance. 'Horace' needed the intervention of his O.D. padre to get to see his mother in hospital. Our leader did not overly like or trust any of us and the feeling was reciprocated. My friend 'George' Hanchett was an even greater thorn in the side of authority, and soon after this he and 'Scouse' Fergusen were both released from the apprentices 'at his own request' as they put it, so I presume that the recognised theory of the time was that it was he who was the bad influence on me. This was in fact a loss to the R.A.F. because both were well above average in technical ability, but just not prepared to take too much of the hard-discipline-for-discipline's-sake line. Had the service realised it, these were precisely the analytical type of tradesmen which were going to be needed in an increasingly technological era.

SHALL OLD ACQUAINTANCE BE FORGOT

Swing low, sweet chariot

During this time we had our final engineering drawing examination at schools. With the teaching standard as it was, all those with previous experience in the subject passed, and most of those without failed. The deeper implication was that those who had failed could no longer receive their ordinary national certificate on graduation, and therefore the remaining two years of schools were largely a waste of time to them, regardless of how well they subsequently did. This happened to me, although I passed all my final schools examinations comfortably, and it eventually caused me to have to repeat my final o.n.c. year at civilian night school. Even the real tryers now became totally disillusioned, and, as a result, class discipline simply went out of the window. I remember the occasion when a piece of cotton was tied round a lamp flex in our class and the end held by one of the electricians behind his desk. When the Flying Officer teacher arrived and the lesson was underway, he began to hear giggles and, turning from the blackboard, he noticed that Pat Cropley was apparently pulling on a piece of something because the lamp was swinging back and forth to the amusement of the whole class. Pat had a very expressive face and was obviously enjoying the whole episode, and making no attempt whatsoever to conceal his hands. The officer rushed down the aisle and snatched at the thread only to find that there wasn't one, at least one being held by Pat.

So it was through a long series of escalating incidents like this that my friend began to take my place in the defaulters' parades. Only two special incidents stand out from my punishment time and both involved an old friend. The rest was just a long unrelenting grind.

One evening I found myself being reacquainted with my old friend, regiment sergeant 'Jim'. He obviously recalled quite well my humanitarian retreat at summer camp, and so it was that after the last parade of the evening I was detained when everyone else was dismissed. Obviously he had a small private matter that he wished to mull over with me, well away from the prying ears of the general mass.

"I remember yew funny bugger," he said in his whimsical way. "Now yew can go and get my beddeeing from my rewm and breeng it eeyer."

This meant a late night return walk with a bulky load for the best part of a mile. It was a windy, rainy night as it happened, and on the way back with the load I slipped, and the bedding went into a puddle. I had no wish to incur his further wrath that particular night, so on arrival I quickly parcelled the blankets and sheets up with the dampest in the middle, hoping to be long gone before he returned to the Duty N.C.O.'s room where he was going to sleep. I was unfortunately unlucky on this occasion, and, when he duly arrived before I had the opportunity to retire, he could not deny himself one last gloat.

"Hah Hah, Breestow," he chortled. "Thees time the joke ers been orn yew. Now we are eevern."

That was a shade too much and caught me on a rather raw edge at the ragged end of another generally bad day. I wasn't able to resist the temptation, come what may. "I'm afraid not, Sergeant. Your bedding has suffered what was, I can honestly assure you, an accident on the way here and the score is in fact 2-1." Assurances or not, it proved sufficient to get me another five days' jankers the following afternoon.

With such a long continuous period of defaulters, it was unavoidable that I saw the complete Duty N.C.O.s list through several repeat cycles, and not a few, on sight of me standing there once more before them in my finery commented, "What again!" only to discover that it was in fact an ongoing continuum with no 'again' involved at all.

It was certain therefore that my friend and I would meet each other once more, and sure enough there he was, and raring to get even, one Sunday. This time I found myself alone in the sergeants' mess with the task of polishing the ante-room floor and the whole of the day in which to do it. It was a very large floor and I decided to be extremely diligent, on this one occasion. The wax was applied and brought to a high gloss with a vengeance and then some, with only the occasional patch of pure grease placed at strategic intervals. Not being the owner of a fine intellect, my friend did not see any great significance in this when he later came to inspect my handiwork before releasing me. He and others did soon afterwards when they began to end up on their backsides unexpectedly. The score had moved to 3-2 but he hadn't noticed the goal.

Sadly, all good things must eventually come to an end, and when I finally completed my punishment and returned to the comparative ease of normal off duty hours, I had been on defaulters for so long that my jacket sleeve was faded where the white armband had been for so long, and so I was still readily identifiable as an old lag and went on being singled out for more of the random punishment. During my punishment period, the Fiftieth raided our rooms one evening and I returned from yet another inspection to find the room in complete chaos. Apart, that is, from my own bed space and kit, which had not been touched. They had recognised that I had apparently quite enough troubles to handle already.

Pat's progress through the ranks of the criminal and rebellious was at a rather brisk pace, and it wasn't long before he was incarcerated in the guardroom under close arrest, as I had been before him. This was as a result of his, shall we say, taking a rather poor view of the failure of a member of the Fifty-Seventh entry to shout, "Up the Fifty-Fifth!" sufficiently loudly as directed, after a small raid on their rooms. Like myself, he was put in the exercise yard alone in the evenings with the door locked while the duty policeman had his feet up. Pat, with ever an eye for opportunity and well informed by myself concerning the procedures of close confinement, hid a towel in his battledress blouse one evening and used it to cover the broken glass on the top of the wall, after he had climbed on the shoulders of a fellow inmate. He had, as they say in the parlance of the gangster films, 'crashed out'. He made his way to number three wing 'tank' and spent a very convivial evening with his friends, drinking tea and listening to the sounds of the mass search going on outside before returning under his own steam to the now thoroughly aroused guardroom and gaining admittance by knocking on the door. You can imagine how popular he had made himself with the R.A.F. policemen required to guard him for the remainder of his happy stay with them.

He was awarded seven days' detention for his multiple crimes. 'Ig' Noble was on booking out duty one evening in the guardroom where Pat was incarcerated under the loving care of Sergeant 'Plushbum' Scott, the 'snoop'. 'Ig' was able to carry back the news of what transpired on this occasion to the rest of us. Pat had apparently been given the pleasant task of polishing the floor under the supervising gaze of this gentleman it seemed, who enquired in light conversation as the work progressed if it was true that the apprentices had nicknames for almost all of the staff. Pat, on his knees on the floor, cautiously replied in the affirmative.

"What is mine then, Cropley?" continued our friend after a while, probably believing that it might be something as innocuous as 'Haggis' or 'Flying' Scott or perhaps even better 'Devil' Scott or something of that sort, but Pat wisely refrained from satisfying the repeated requests for such an exchange of confidentiality for obvious reasons. Finally a warm assurance was given that there would of course be no hard feelings afterwards. Pat was never so foolishly gullible as to believe that, and had probably been pondering for some time that he might well be far better off locked up but at least resting in his cell rather than carrying out this sort of unpleasant manual hard labour, so finally he told him.

"It's Plushbum," he said. He was of course quite correct in his assumption and was back in that cell in minutes.

Shades of 'Ianto' and friends and the battered 'snoops' returned to their minds one evening when 'Ig' and a companion were taking an evening stroll in the woods and noticed that they were apparently being followed by two such white-capped gentlemen, who obviously believed that the apprentices were off for a spot of illicit drinking somewhere. Accordingly they quickened their pace, took strange diversions, ran, hid and reappeared and generally had a high old game for a couple of hours until the 'snoops' gave up and retired almost exhausted to their guardroom home.

Notoriety tends to attract those who, while being of a like mind, need just that little inspirational push, and Pat soon had a loyal supportive Irish engine fitter friend called Paddy Fitzgerald in almost constant attendance. I believe that the modern term is a groupie. He was a little imp of a fellow, and the absolute epitome of the minuscule Irishman who is always backing up his larger muscular friend with tactful calls like, "Yers is not goin' to let him talk to yers like dat are yer Paddy?" Pat thankfully did not respond to the dangers of this kind of hero worship. At that time he was capable of getting into quite enough trouble with absolutely no assistance at all, thank you very much, but he was indulgent and kindly with his small friend and began to call him 'Fritz' Gerald for some unknown reason, and soon so did everyone else, of course.

Fritz, like myself, was not enjoying an easy apprenticeship. In his first end of year exams he had been required to write all that he knew about the measuring device called the micrometer. All that he could recall was that it looked like a small wrench, so he had begun with a statement that it had the appearance of a finely adjustable spanner. He could remember little else of significance and was staring glumly at this rather bald statement, sucking his pencil and awaiting inspiration or divine intervention, when the class supervisor noticed his inactivity and came and read the very little Fritz had produced over his shoulder. He sarcastically enquired if that was all. Fritz quickly added, "And can be used as a hammer," in his best handwriting. He was soon on a charge for this little piece of witticism. Whilst we only have Fritz's word for the truth of what subsequently passed in the C.O.'s office, I myself believe wholeheartedly in the accuracy of his version. It seems that, when that certain vital piece of written evidence was produced for the officer's inspection, he read it and then asked of the witness, "By the way, can it?" So it was that poor Fritz was given three days' punishment for not knowing what his C.O. did not know. A sure and certain sign of the general democracy being shared by all.

There had been some changes in our ranks over a period of time too, mostly during the second and third years. Billy Broughton, my music teacher, after a long sickness was finally moved back to the Fifty-Sixth entry. Four former Fifty-Fourth entry member electricians came down to us in the fullness of time. Leading Apprentice Tony 'Rowey' Rowe, and 'Swede' Walters first, followed by 'Necklace Mo' Critchley and 'Ginger' Saunders, and two of our original members, Curwyn 'One Ball' Lewis and Clive 'Ginger' Evans, had already been given earlier medical discharges for physical defects that the Air Force apparently couldn't fix, although our regiment sergeant friend 'Jim' had assured us daily that the Air Force could do anything at all, including getting us pregnant, with the one single provision that it could not then make us love the child. 'One Ball' had managed to regrow the hair round his bottom by the time that he departed, you will be happy to know, so he did not re-enter civvy street quite naked. Such was the unity engendered by sharing the common experience of the first year that people invariably retained their image of belonging to their original entry regardless of such moves. It was this close identity and entry spirit that was and always had been a persistent thorn in the flesh of the 'they' and which, in my opinion, had prompted them to split up our entry into acknowledged 'goodies' and 'baddies' as a preliminary move. Later they even divided up all the wings by trade in a further ongoing purge, but the die had already been well and truly cast by then, and it completely failed to change anything at all.

As a new baddies squadron we now had Dick Corser to contend with all over again, back as our disciplinary sergeant after a short absence. Previous experience at least suggested that we should perhaps not entertain too high hopes that there had taken place some great and significant character change and spiritual uplifting in Sergeant Corser during this absence, and earlier dreams that we might perhaps be fortunate enough to inherit someone a little easier now proved to be totally unfounded. This left our old friend Flight Sergeant Reid in charge of the old original Fifty-Fifth squadron. He had long been known to us all as 'Red Eye Reid' from a common belief that he suffered from a permanent eye infection due to the frequent ongoing application of his orb to any keyhole in an effort to discover evil taking place within our midst and to ruthlessly eliminate it.

On one occasion he and our good friend the regiment sergeant thought up a jolly good wheeze for the electricians after yet another small infraction of the rules. After forming the whole lot of us, they set us all off marching on the parade ground and the adjoining road under the command of our very own 'Ig' Noble. This was just for openers. They then moved 'Ig' into our ranks and enjoyed a contest to see which of them had the loudest shouting voice which could allow all of us to march some considerable distance away from them, and still hear and of course obey the order to about turn and march back to them. Once we had been up and down that road a few times and discovered the exact nature of this amusing game, it didn't take us long to bring it to a sudden and unexpected climax. We were at some considerable distance away when it was the turn of 'Red Eye' to bellow out his order. All of us became suddenly afflicted with acute deafness and carried straight on. He opened his mouth, increased the decibels a notch or so and tried again but with no success, sad to say. Regiment Sergeant 'Jim' then smugly intervened, plainly believing that the bet, whatever it was, was now most surely his with an even louder shout. By this time the distance was so great that we in the mass were in fact disappearing over the brow of a hill in the road, and we kept straight on as before and disappeared from their sight. Then of course a certain degree of panic set in, and the pair of them were obliged to break into a brisk run in order to catch us up before we marched off the end of the world as our discipline surely had taught us to do. We did rather better than that, however. Once sure that we were at last completely out of their sight, we had voluntarily broken into nothing less than a good full gallop until we reached the safety of our barrack block, into which we disappeared. It must have been a joy to see their faces when they reached the crest of the hill and found us gone. We didn't play that particular game again for some reason.

Pat, 'Dai' and I were now housed in a room of airframe fitters including 'Skinful' Stentiford, Dick Pearman, George Guttridge and 'Flaps' Picken, with Ted Patterson sleeping on one side of me and 'Waxy' Crane on the other. 'Dai', a typical Welshman with a great love of singing, and recognising that these were simple rude mechanical types of people, unused to the company of electrical gentlemen, set about their education by teaching them to sing in Welsh as he had done from the earliest days with his electrical comrades. I know for a fact that at least one of them can still sing "Wenn, Wenn, Wenn", to this very day, as I can myself. Ted was a Scot with that often found characteristic of being a trifle chatty, to put it mildly. He had accordingly become nicknamed and known to all as 'Mighty Mouth', a corruption of the rodent cartoon character of that time. 'Waxy' had been an original member of the room which had been shared with the overspill electricians, and, no doubt corrupted by the contact, had long been accepted by us as almost a fellow tradesman. He was also by now an accomplished side drummer in the pipe band, and many nights spent listening to him tap out the drum rhythms on absolutely anything that presented itself have locked the patterns into my head permanently. Perhaps it also accounts for the headaches, and the odd occasional moment of erratic behaviour, but 'Waxy' was and is absolutely one hundred per cent reliable. Like many others whose homes were within hitchhiking distance of Halton, he was generous to those of us who were not so fortunate, and several of us had enjoyed his mother's hospitality for a whole weekend, sometimes in a group. I had also been home with Pete Deverall and Pat Cropley, visited

Paddy 'Swing Ring' Swoffer's grandmother in elegant Cheltenham on the back of his illegal motorcycle, and 'Horace's' farming cousins in Newbury. We had learned by then to share almost anything it seemed, including relatives.

I recall standing at attention next to 'Waxy' on one Saturday morning room inspection, when for once it was he rather than myself who managed to attract the attention of Dick Corser. After failing to detect any irregularity in 'Waxy's' appearance, although he really did try his most unpleasant best, he finally asked 'Waxy' to hold out his hands and then complained that his nails were too long. That was high praise indeed in those days, although I doubt that 'Waxy' managed to see it that way.

'Baddies' we most certainly were, however, and we were only too pleased by this time to confirm the opinion obviously held of by our masters should the situation present itself. It did one day when, from the windows of our room, a group of us observed that first the grass and then some of the wooden huts located behind our barrack block had caught fire, judging by the amount of smoke pouring from them. This happened to occur at a most auspicious time, since not too long prior to that we had in fact received the benefit of a lesson from no less than the station fire fighting department on one of our G.S.T. afternoons.

The flight sergeant in charge of that fine body of men had personally demonstrated just how the man on the fire hydrant was first required to call out, "Water on!" to which the man on the end of the hose separating them was required to repeat, "Water on!" before any action should be taken; and when the fire was extinguished the call was to be, "Water off!" and the acknowledging reply the same. Of course it did not escape our youthful notice that his cry was rather more like "Wowarorn" and "Wowarorf" and accordingly, when ordered to make the appropriate cry ourselves, we gave rather better than passable imitations which he apparently interpreted as being highly complimentary. There was a certain amount of worthless trivia sandwiched between these two apparently very significant shouts which we soon forgot about because unfortunately the instructor had been so enthusiastic about the whole lesson that he had chosen to play both the hydrant and the hose men's parts himself, which seemed to require a great deal of nipping about from one end to the other, taking up both positions in turn in order to give us the true in-depth feeling, with the appropriate shout of command followed by the run to the other end to give the required reply and then back again. As our incredulous eyes had followed his ungainly sprints up and down before us, we had not been slow to notice that a distinctly amusing possibility was emerging, and when question time arose it was not long before we had him running up and down all over again to the point of exhaustion, and shouting his "Wowarorn" and "Wowarorf" orders at our impish will. It had all been most enjoyable and now fate had chanced to deliver a practical application of that very lesson into our eager young fingers.

The wail of the approaching fire engine's siren could already be distinctly heard in the distance so the moment of truth was really at hand. As good responsible young airmen, we rushed down there to the scene at once of course, but, I must add, without the weight of any of the barrack block fire extinguishers to burden us, and we hurriedly blocked the access road with dustbins, rocks and anything else we could find lying around to prevent the fire engines getting anyway near the fire, before returning to the vantage point of our overlooking window to watch the fun. Judging by the amount of last minute frantic braking, shouted recriminations between the firemen, swearing and arm waving that went on, we were most successful on this particular occasion I am happy to say, because at least one hut burned to the ground before our eyes. As we watched the unsuccessful fire engines load on their crews and finally depart the still smoking ashes 'Waxy' called after them anxiously from our window with a parting admonishment, "Wowaorf! Don't forget now. Wowaorf."

During my time at Halton, I only once applied for any privilege, but the occasion that I did so was for a very special reason indeed. My father had written to say that he was coming to London on a football trip to see Grimsby Town play the Arsenal in a First Division League match, and I was granted a late pass for the day to join him. Our common love of football has, by the way, survived two more generations since then, and for me it was one of the very few times that we were close together and sharing a common interest. Unfortunately it was also the last match that Grimsby Town were to play in the First Division and they lost comfortably that day too, but, standing on the terraces and having a drink together afterwards, it was nice to feel that after all I was his son, and now almost a man myself. I never forgot it, and surprisingly neither did he.

Our good electrical friend who, you may recall, had taken up an early membership in the radio club for doubtful purposes, had been coming along technically at a pace in the past year and a half or so. In the fullness of time, he constructed a remarkable device – a spark plug mounted in a seven pound empty jam tin as an antenna – which he fixed on the roof of the radio hut which was located just down road from the sergeants' mess, and he connected it to a hand-turned magneto located in the hut itself, with the sole purpose of interfering with the sergeants' mess's recently purchased and latest ultra-new innovation, a television set. Then it was purely a matter of looking up the broadcasting times of likely programmes of interest in the newspaper in the N.A.A.F.I., leaving a suitable time interval for the generation of some genuine programme interest amongst the senior N.C.O. audience and then just starting to crank. We knew that he must be enjoying some degree of success in his experiments when the grapevine informed us that no less a personage than the Flight Sergeant electrician himself, at the request of the Mess Committee, had begun to carry out a personal check on all the electrical appliances in the cookhouse, which was located next door, looking for possible sources of interference. There was never a lack of volunteers to crank the magneto after that.

'Flaps' and I became firm friends over the next few months, and he eventually took me home with him to Lowestoft for one of our leaves. Matters did get a trifle out of hand when we both began to fancy and actively pursue the same young lady, and I felt that as my host the simple laws of hospitality dictated that he should defer to me in such pleasures, and additionally, if further rule was needed, the last three digits of my service number were 602 and certainly lower than his, which clearly made me his senior. He unfortunately chose to see it otherwise. However, our entry comradeship apparently extended beyond such mundane matters as women, so for the rest of the time we shared her, going everywhere in a trio of two happy males and one completely bewildered female, and when the leave finally ended she saw us both off with kisses from the railway station.

Other patterns locked themselves into place in my head, and into the heads of my fellow apprentices at that time. On ceremonial parades, sometimes standing at attention for hours while some senior officer slowly inspected every rank and file, one needed to permit the mind to detach itself and locate and fix upon some remote comforting thought, far away from the actual physical discomfort being endured. Although I did not appreciate it at the time, this was in fact excellent training for a future period in my life, when, unable to invent a handy scrounge, I would sometimes be obliged to accompany my wife on one of her clothes shopping trips, and be required to stand in the middle of the ladies' underwear department staring into the far distance, apparently oblivious to all that surrounded me and of course interminably waiting. After Halton it was all too easy. In those early days the military and pipe bands, playing the by now familiar slow marches certainly helped, but even better was silently singing the vulgar versions of the words of some of those marches. If you happened to glance around out of the corner of your eye on one of these occasions, you could see perhaps ten or eleven pairs of lips each silently mouthing the same, "Somebody shit on the door step." It was invariably a great comforter.

A FINE MORAL UPSTANDING EXAMPLE

Ring the bell verger
Ring the bell ring

At home on summer leave again and fast growing up, I was now not only sufficiently independent but had time to look around me and reflect again on the situation that had prompted me to leave home in the first place, eighteen months earlier, with a softer more mature eye. I began to belatedly realise that my father had in fact made some very fortunate implicit decisions when he had decided to remarry. He now had a family around him again, and that was the place where he truly flourished as a human being, in his own home. My step-sister's son Mike had virtually become a new son to him, and had obviously comfortably filled my missing slot. The home was cosy and my stepmother looked after my father very well. They had become a new whole, but, of which I realised every time that I was on leave, I was no longer a part, and I began to concede that, since I would have eventually left home anyway, he had done the best thing that he possibly could. I was at last coming round to taking a kinder view of what had gone before, although now his vision of what constituted his family was no longer mine. Although father and son, we were essentially very different kinds of men, you see, each with our own individual strengths and weaknesses. I began to realise that if I was the one to have been given a larger imagination, then it came with an implicit responsibility for me to use it and have some understanding too. You simply cannot condemn someone for the rest of your life just because, at one point in time, they happened to lack what you did not. Sometimes in later years, I will admit that it would still hurt when he called me Mike by mistake, but it was for him as a man to build his life and to accept what came with it, and for me to do the same.

Leave itself was not all fun for me then, even though it was a most welcome respite from the apprentice's lifestyle. My old friends from civilian times were now rapidly becoming just that – old friends. At work all day, they now had their own less urgent leisure pattern and new friends. Their interests and experiences were no longer mine, and I didn't seem to fit in any more. There was rarely a subject of common interest between us, apart from girls of course. The never being a prophet in your own country was an accurate observation I found out. I was still clumsy and ill at ease with girls in general despite the interest, and my monastic apprentice's lifestyle had done absolutely nothing to make that any easier. On one of my leaves in that final year I met an old junior schoolfriend Harry Prior, who was himself in the army apprentices and stationed at Arborfield, and, sharing the same feelings as myself about fitting in, we spent our leave days together.

Harry was an opportunist par excellence, who had discovered that it was possible to win at will on some of the amusements at the local Wonderland Pleasure Park, close to my home, and to sell the prizes to the seasonal holidaymakers from the industrial Midlands who had less talent than himself, in order to subsidise his meagre pay. The particular attractions in which he specialised were ones in which the punters were required to roll two rubber balls up a sloping table, where they fell down through a series of numbered holes before being returned to you, and which clicked a wooden horse forward according to that number, in a race against the other player's horses. Harry's method was simplicity refined. He just brought three more rubber balls with him, so increasing his throughput, and he invariably won, until he was banned, that is. From then on he was only allowed to take part if trade was slack, and the fairground barkers needed someone in uniform to bring in the crowds. It was still just after the war you see, and servicemen still enjoyed a certain air of glamorous euphoria about them. He would then get the nod that he could take part. In less auspicious times he applied his initiative to assisting the dodgem car attraction, when in civilian clothes, of course, not wishing to bring his uniform into disrepute, by springing nimbly on to the back of the cars and taking the money from the intending drivers, just as the employed assistants did. Of course he didn't actually share the proceeds with the owners but that is a small matter. The difficulty for me with being in Harry's company was that soon I was also banned, merely from being seen with him. That all changed one night when we were both watching the happy holidaymakers crashing into each other in the electric cars when suddenly the whole dodgem arena went dead. It was driven by a large mercury arc rectifier which had tripped out on overload, and no one in attendance apparently knew how to restart it. Fortunately I had been awake for once when that subject had been taught barely weeks before in the Halton classroom, and so I offered my highly skilful services, at a price of course. From then on free rides and friendly smiles replaced the banning, at least for the period of that leave.

Apart from this, I kept company with 'Goody' Goodacre, an instrument-maker from my own entry, when on leave, although, in all honesty, I barely knew him when we were at Halton. He never as much as did a single day's defaulters in his whole three years, so we did not exactly move in quite the same social circles in working time you might say. Unlike myself, he was always most anxious to be helpful and to prove his worth and that tended to get to the point of almost getting him quite unwittingly into mischief on occasion. He was, for example, once in a workshops class where the invariably irritable instructor was asking probing questions of some poor instrument-maker or other who was experiencing some difficulties in finding a suitable answer. In jumped 'Goody'.

"I know! I know!" he interrupted. He was told to remain quiet and wait his turn whilst the inquisition continued but no, he just had to go on. I could have advised him, having learned almost two years earlier from my own experiences, that it was prudent to volunteer absolutely nothing, but, there you are, some people are just naturally slow learners. "But this is right up my street, sergeant," he persisted with enthusiasm. All of the instructors, regardless of trade, had been hand selected in the first place, in my humble opinion, from those people with the shortest possible fuse on their tempers, and daily association with us apprentices of all entries had pared even that down to a very fine hair trigger indeed.

"Shut up!" exploded the sergeant, waving his wooden blackboard pointer about wildly in all directions, "or this will be up your bloody street Goodacre!" The news of this charming interlude spread through the barrack rooms that evening.

'Goody' and his two male civilian cousins used to go out on Saturday evenings when on leave for a quiet drink, at which I was already tolerably good, and dancing, at which I was most certainly not, in joint company. My belated introduction and attachment to 'Goody' came about in rather extraordinary circumstances and was the result of his experiencing some problems with his mother at the time. His only crime was that his hormones had rather horrendously woken up, and he had got letters intended respectively for his current girlfriend and his parent inadvertently into the wrong envelopes. I cannot know for sure quite what it was that he expressed and set down in ink on paper as his intention on the next leave, but his widowed mother was, shall we say, not entirely pleased with the obvious current state of his morality, in fact she was so considerably alarmed as to seek advice and assistance. Her sister happened to be my father's next door neighbour, who had known, or at least thought that she had known, me from my happy toddler childhood days. 'Had known' being the key phrase, since much had happened in my life of late, and the polite little boy in velvet trousers playing in the garden with his bucket and spade that she remembered was sadly long gone. The two ladies jointly came to believe that I might just be the very one to become a good influence on 'Goody' and get him back on the straight and narrow, and so they encouraged us to go out together. Which just goes to show what value one can put on advice which comes from within one's own family. Had they known of my reputation and current standing as regards lawlessness in the whole of the Fifty-Fifth entry, they might well have changed their minds. He and I both found this heartily amusing and divided the responsibilities equally in good apprentice fashion. We departed from his house in joint safe company in the evening and returned the same way, and, whilst I then took care of the drinking, he pursued his evil way with the girlfriend and everyone came out happy.

This was a time of a total growing awareness, and one learned other basic lessons about life even when away from Halton. You can never completely leave home, it seems, for example, and are forever feeling out your roots given the opportunity. One day, when on leave, I happened to be close to Victor Street, where I had been born. I suppose that I was retracing my childhood and my Saturday morning shopping trips with my mother without even realising it. After all, it was still only three years since her death. I suddenly took it into my head to visit my step-grandfather, although I had not exchanged words with him since well before her funeral. Much had happened over the intervening years, and I decided that perhaps now might be a suitable time for both of us to talk about things that we perhaps now could discuss as something approaching equals. His own story of how he came to marry a widow from the First World War, with seven children, and all from a pen friendship, was remarkable enough in itself.

Things had most certainly changed, even judging from a casual glance. Part of the street was already demolished in a slum clearance plan, including the ice factory and it plainly wasn't going to be long before his house, where I had been born, went too. Perhaps that is why they had failed to erect a brass plate to commemorate such a significant event, I considered. When I visited the place again thirty more years on, what had been under construction the first time was now also being demolished in turn, and again in a slum clearance plan, which made me speculate that the brass plate wouldn't have survived for long anyway. The lady who answered the door to my knock was a complete stranger to me, and she told me with shock that my step-grandfather had already been dead for over two years. No one in the family seemed to know anything about it when I asked at home, so it seems that it is entirely possible to not only live alone – and we all certainly die alone – but that your passing may still be totally unrecognised even if you choose to live in the same small-town close community all your life.

FLASHES OF INSPIRATION

If the circuit don't trip
You'll have no balls at all

Back at Halton again, significant events were on hand in Maitland wing. The passing out of the Fiftieth entry had been expected to be spectacular. One sergeant apprentice from a junior entry had needed to be moved off camp for his own safety. My friend the P.T. instructor came in for the very special attention of their electricians. They had also been on the receiving end of some of his injustices, and so on passing out night they set a booby trap when they learned that he was to be the duty N.C.O. that night. Their intentions were well known amongst all the electricians of all entries, and relied heavily on a piece of basic psychology. They opened up the fuse box in the hall of the barrack block and bridged the fuse with a large flat strip of copper, then left the box deliberately and obviously wide open. Sure enough, in his flashlight checks the sergeant saw the open box and the copper strip, and sure enough, with predictable foolishness, he decided to remove it. I saw the flash from my bed several blocks away, and a shiver went down my spine for him that night.

At the passing out parade, while we waited our turn to march on, the two apprentices chosen to be the standard bearers and the inspecting officers escorts stood nearby. They were chosen for their tallness, general bearing, and smartness, and in the Fiftieth that was something. One of them, Corporal Apprentice 'Gannet' Gross, so called because of his enormous appetite, was stood resplendent in white gauntlet gloves beside his pennant standard. One of our entry cheekily called out to him, "Where are your eating irons now, Gannet?" With a great show of disdain, 'Gannet' rolled back the cuff of one gauntlet to reveal the knife, fork and spoon that had never been far from him for his three years.

I did not see for myself but I did hear of a miracle which took place amongst a group of our engine fitters round about that time. Being now almost eighteen years old, they had developed some rather sophisticated tastes, such as beer. Forbidden as it was, and with the local pubs being unwilling to serve you, it was a matter of pride to overcome such minor obstacles, of course. After lights out, therefore, two individuals would sometimes, after a pay day, climb down the drainpipe carrying the water bottles of the thirsty and climb the Chiltern Hills behind our barracks in search of small village pubs with low cashflows and weak moral fortitude. One weekend evening they were either careless or unlucky because they were caught red-handed trying to get back into their barrack block. The water bottles were taken as evidence and locked in the C.O.'s safe for safe keeping overnight. It chanced that one of their room was the possessor of some unusual skills and in fact claimed that he had learned his craft from his father who was a burglar. Prudence forbids that I write his name although I remember it quite clearly, since in the austerity of the present day world he may well have had to fall back on those skills to provide sustenance. With some assistance in providing light from a torch, he then demonstrated that it was not all just talk, as he first picked the door lock and got into the C.O.'s office. He then displayed his real talents by easily getting the safe open, and that is where the miracle of changing the bitter into water took place. The following day before the C.O. the defence was presented that the two had merely been guilty of being absent from their beds after lights out, and obtaining spring water from one of the natural wells found in the hills for athletic training purposes. The safe was opened to obtain confirmation, the contents tasted and the charge was dismissed.

The matter of strong drink tended to be handled by trades in their own individual ways. The armourers, for example, were reputed to be operating their own exclusively in-trade still. 'Ig' and Robin combined their resources to attempt to make elderberry wine in the bath, which temporarily made it off limits for the use of the rook entries of which they were now in charge. In a haste to accelerate the fermentation process before discovery and under the rising tide of complaint from those who wished to remove the daily grime from their bodies, the product was bottled rather early and stored under Robin's bed in the privacy of his room. The bottles unfortunately blew out their stoppers and both he and 'Ig' were urgently required to clean up the mess and get rid of the smell. Inspired by a tall story from an instructor concerning his life of adventure in far off sunny climes, their next attempt was to drain the de-icing fluid from a

Mosquito aircraft at the airfield and mix it with still lemonade. It made both of them violently sick, and at that point their further experiments ceased.

I was myself more circumspect in my dealings with the wing people in those days. We had a new disciplinary sergeant, 'Tug' Wilson, who now translated the wishes of our still morose commanding officer into positive and usually unpleasant action. 'Tug' had some badly fitting dentures and the sounds that emerged, especially when in anger, were not always entirely intelligible. He arrived one afternoon in our room with the intention of getting us changed into P.T. kit and outside in double quick time. It began when Tony Large suddenly became rather deaf and asked the sergeant to repeat his order. He foolishly obliged and the teeth went into a rapid clickety-clack. Everyone then failed to understand and it grew worse. It was then that I came to his rescue.

"What he is saying is that we are to wait here," I interpreted. Clickety clack went the teeth. "I'm sorry," I apologised. "What the sergeant wants is that we put on our overalls and fall in downstairs." He was desperate and becoming increasingly angry, but he only had me to help him at all. You would be surprised how long this went on and even more surprisingly, he obviously liked me when he finally gave up and left. My roommates wryly accused me of starting to be a creeper.

'Tug', in his anxiety to ingratiate himself by pleasing absolutely anyone in authority, was on occasion apt to make himself singularly vulnerable when he least expected it. He had been giving the electricians a spot of extra drill for some crime real or imaginary one day, when on to the parade ground ran a small dog which took up a position at the side of 'Tug' and sat down facing us. We looked at both and grinned. 'Tug' looked at the animal and scowled. The dog looked at 'Tug' and I swear that it smiled. It was immediately recognisable to us brats that its owner must be of service origins since it clearly not only knew how to recognise exactly who the enemy was, but what was to be done with us too. As 'Tug' bellowed out his orders, it joined in and began to bark at us in unison, in much the same tones that 'Harold' normally used, and was in fact fractionally more understandable than 'Tug' himself, come to think of it. He was not at all pleased with this, nor by our barely concealed smiles, and, after issuing a few furtive growling noises out of the corner of his mouth, which only served to make the dog bark at him as well in protest, he finally lashed out at it with his ammunition boots. The dog, although small, was no timid creature however and stood its ground well until 'Tug', finally losing his temper, began to chase it, waving his pace stick in the air like a cavalry sabre, taking a few lusty swipes at its hindquarters and emitting bloodthirsty shouts. At that point, out from the Wing Headquarters rushed 'Steve' with shoulders rising and fingers curling and uncurling in very rapid tempo indeed and shouting for all to hear.

"Sergeant Wilson! What on earth do you think that you are doing to my dog." 'Tug', thoroughly nonplussed, didn't know quite what to do and began making bleating noises of, "Sah! But Sah!" which did nothing at all to placate 'Steve' who, after a final parting glare, picked up the poor animal, which was clearly very happy to be reunited with its owner, and they returned whence 'Steve' had just come, with his making most uncharacteristic childlike soothing noises to the animal. The drill period ended rather earlier than usual on that occasion, but the more brave of us and those with a taste for risk taking took to making barking and growling noises outside 'Tug's' office for the next few weeks, although one always had to be prepared for an early retreat should the door suddenly open.

At the C.O.'s suggestion 'Tug', ever anxious to please, yet again managed to get the cart before the horse, and he set up an evening study facility for us in our final year, located in a previously empty room on the ground floor in our barrack block, and provided with coloured posters, tables and chairs, manuals, the lot. It was of course brought to a very high standard of cleanliness by the customary application of delinquent apprentices' sweat and muscle, and then remained securely locked and only opened for C.O.'s inspections or further bulling sessions.

One Saturday morning, with the snow ankle deep on the ground and the wind bitingly cold, our entry was suddenly assembled outside, clad only in P.T. kit, for a cross country run. It was clear that at that hour we would not be back in time for lunch, or for the ones who lived quite near to be able to go home for the day, so another small sickener was intended as our lot. I for once had been lucky, and had been sent to the cookhouse to clean greasy pans for some small infraction or other. The lowly 'boggie' detailed to supervise my efforts, as I recall, had decided that this was indeed a most auspicious occasion on which to start to play the hard man, and, after tolerating this for a remarkably protracted period of time, a tribute to my infinite understanding of my fellow man and sound service training, even I had finally run out of patience. I felt obliged to remove the hand chopper which was set in the butcher's wooden block which I was scrubbing at the time, and to advance on him with it grasped firmly in my right fist and solemnly advise him to depart elsewhere and do disgusting things to himself, or alternatively get used to speaking with a high pitched voice. From that simple chance caring remark, would you believe it, our relationship suddenly blossomed, and he was not long too in warming to my general personality, and even going so far as to release me from further irksome duties before the allocated time. There was still no sign of the return of my friends, however, and even I began to be anxious concerning their fate in those arctic conditions.

Then, coming up the hill in ragged order, I spotted first Pat and then a few more of our good friends all smiling and splendidly filthy from head to toe with coal dust. The rest of the wing, officers, N.C.O.s and all, watched with wide eyes as this small group trooped in to the barrack blocks. It seemed that, after reaching some far distant turning point, the flight sergeant P.T. instructor, none other than our good friend 'The Muscle Man', had declared that it was now every man for himself, and that the slowest would have to fend for themselves and get back as best they could in their own time. As soon as he and the others were out of sight, Pat and friends had done just that, and thumbed a lift on the back of a passing coal cart. It was much too cold to sit out there unprotected so they had covered themselves with the tarpaulin normally used to cover the coal sacks. You can guess what they looked like. They were followed soon after by a second group who had stood at a bus stop in the shivering cold in their shorts and vest, flagged down the next passing bus and explained that they were on an initiative test and that therefore they had no money for fares. The kind-hearted ex-serviceman driver, no doubt reliving some of his own unfortunate service experiences, had obliged them, and the old lady passengers on their morning shopping trip had declared it to be an absolute scandal, sending young boys out in those conditions and almost without clothes. The last one back was the flight sergeant himself, abandoned over the last mile or so by the younger more fleet-footed of his charges.

My friend the regiment sergeant had managed to impart to me something of the art of guile and low animal cunning practised by the N.C.O. s over the previous two years, and I was by now ever alert to the possibilities of a sudden unexpected confrontation. The door of our barrack room burst open just after lunch one sports afternoon and in he strode. He never did manage to master the subtle trick of turning the knob and pushing gently against the door.

"Yew yew and yew get your rifles and come with me," he bellowed. "I erv a treat for yew. Yew are all goin on the rifle range."

Being one of the members of his privileged inner circle of close acquaintances, I was naturally selected to be one of the 'yew' of course, and my brain immediately changed into rapid high gear. My rifle meant dirty sling, in which matter I had unfortunately been somewhat negligent of late. Punishment number one seemed likely. Range meant shooting, equals rifle to be cleaned for many days afterwards as the barrel sweats, equals intended multi after-checks and possible eventual further punishment. My overall assessment of the situation was that it was probably another small sickener for a selected few, rather than any treat or training programme for the many as was stated. I rose to the occasion magnificently in true service fashion and took 'Swede' Waters's rifle since on evidence the sling was certainly far cleaner than mine, and he was away in hospital at the time and consequently therefore in no desperate need of a weapon. If further justification was needed, he was, when all was said and done, only ex-Fifty-Fourth and so didn't really count anyway. Once out there the same old monotonous procedure began.

"Five rounds! At your target in front! Wait for it!" 'Jim' shrieked in typical fashion. Where else could the target possibly be, I reasoned, and if elsewhere how did one fire at something behind you? 'Swede' had been in dock for a good deal longer than I had realised, because in the meantime the rifle had not exactly been maintained in good repair, and when I fired the first shot, it kicked like a runaway horse, and hit me in the nose which then began to bleed profusely. I saw the hand of God in that one all right, but I knew better than to make any complaint however, or to request assistance for my grievous wound and so I lay exactly where I was, all bloody-faced and went on firing. The sergeant, strolling down the line of the prone riflemen, noticed my discomfort and the still steady drip of blood, and gave me one of his rare smiles, which could be confused with wind pains if you were not the observant type.

"That's the way, lard," he muttered, "Stay steel like I erv taught yew."

This was a pure bonus to him, I realised afterwards. At the very least I had suffered from having a dirty rifle, and at the best I had almost been driven so far as to attempt to do myself in.

My small brush with the member of the cookhouse staff was not untypical of the times. These rather humble and simple airmen saw all too well how we were being treated, and assumed that it was all right for them to join in too. One of our entry was chatting up a W.R.A.F. airwoman one evening when two such airmen from the nearby School of Cookery passed by and began to give him at first a hard time and finally a good beating up. We held an entry emergency meeting in the 'tank' and decided to curb their obviously growing enthusiasm. We raided their billets that night and wrecked the place from end to end. We also located the two hard cases and set them up for a quick rematch, but this time with our rugby team captain. He laid them both out in no time flat. All this would seem to have been poetic justice, but unfortunately we happened to have chosen the evening before their A.O.C.'s inspection, so he was treated to a full frontal centre page view of the damage the following morning. He didn't seem to see it our way at all, and there was a lengthy period of punishment for us all yet again. We made some impression on our cooking friends however, because one of them happened to be on the same overseas draft as myself some years later, and, on noticing my service number on my kitbag, asked if I was an ex-apprentice and if I had been in number three wing, because he remembered both the incident and my face. I recalled going into one of their rooms, picking up the first wooden foot locker that I came across and dispatching it and its contents through the whole window frame. When I had turned my attention to the bed I found that someone was lying beneath it for shelter, who begged me not to hit him. It was my fellow draftee.

A new privilege came our way with our rising seniority. We were allowed to go to second house pictures, if we had the money that is, which was sixpence in those days. This was far better entertainment than you would ever believe. The film itself was little more than a silent background, with the audience supplying an ongoing spoken script of their own, usually vulgar and with a strong reference to our own daily experiences. I saw *The Jungle Book* at one of these performances and recall the python snake's sinuous movements along the ground as it crept up on its prey in true reptilian fashion, to a sudden loud comment of, "Now exactly who does that remind you of?"

The very last inter-apprentices raid in which we were involved was a complete fiasco. Two of the number one wing entries decided to combine forces and do to number three wing what we in earlier years had done to them in co-operation with the Fiftieth entry. They did manage to get into some of the Fifty-Eighth rooms, but we were well alerted by that time and awaiting their arrival behind erected barriers and booby traps with great interest. If previous form was anything to go by, it would be they who would be required to pay for any damage, and it was highly probable that there was going to be plenty of that. In the event they never made it as far as our rooms, but 'Dai' Evans, my fellow electrician, chanced to be returning from a 'wad and a char' in the 'tank,' and, wondering what all the noise was about around the Fifty-Eighth billets, went to have a look for himself. Suddenly his own commanding officer leapt out of the shadows, threw his arms round 'Dai' and began shouting, "I've got one. Help! Help!"

'Dai', being his normal cheerful self, tried his gentle calm best to explain that he was in fact one of this officer's own men, and not some number one wing hooligan out on the prowl. He should have known better than to even try, and he soon found himself, together with some of the genuine attackers, locked up for safe keeping in number one wing guardroom. Simply no one would believe that he was from number three wing. The following day he was awarded a few days' jankers anyway by this same officer who must have by then realised his mistake.

There were so few opportunities to avoid anything at all that one simply had to seize any small gift when chance placed it in your path, no matter what ethics or principles were involved. I was one of the many electricians who suddenly saw the light and turned to religion when we stumbled across the fact that you could be excused the worst of Friday evening fatigues if you were attending the Padre's confirmation classes that evening. All that was required was an earnest, serious face and a tendency to clasp your hands in front of you with your head bowed. Those of our number who were without those abilities were soon taught them by the less scrupulous and better actors. In view of the average quality of the staple diet being dumped on our plates each meal time, it proved to be no difficult thing to learn to eat the little pieces of blotting paper, and basic fitting hardened fingers soon learned to wrestle wine goblets from strong clutching ordained hands. Mummies and Daddies came to watch no less a personage than the Bishop of Aylesbury himself as our first opponent. I remember that he had a grip on that chalice like a vice and knuckles the size of walnuts from long and distinguished service in this particular field.

At roughly the same time, I also rapidly 'unlearned' to swim when it was announced that all non-swimmers were required to take lessons while the rest were on cross country runs during the miserable winter months. Pat Cropley and I happily thrashed around in the pool, doing our drowning act, with no noticeable improvement whatsoever for the whole of several months. It was Pat's great joy to always put on a great show of reluctance to get into the cold water when we arrived at the pool's edge, and to gingerly climb down the steps for some prolonged toe testing, rather than bravely dive in from the edge, and only then after much shouting and abuse from the physical training instructor in charge. Pat would then stand on the bottom, in the shallow end of course and allow a simply wonderful look of relief to slowly spread across his face. Invariably he was told just how disgusting he was in no uncertain language, and he obviously enjoyed every single moment of it. Personally I always moved well away, just to be on the safe side in case he didn't happen to be kidding. You could never be too sure of Pat. It was finally arranged that the new swimmers would provide a swimming gala to display what had been taught them during this time. We were required to march out in single file, line up at the side of the pool and dive in before doing two brisk lengths in whatever style was easiest, including dog paddle if necessary, the main requirements being two lengths, no detectable foot on the bottom stuff and absolutely no drowning in front of the audience of course. One was supposed to be clad in the regulation dark trunks, that went without saying and that was where Pat and I decided to make it all go wrong. His sister Yvonne had some material left over from a summer dress. It was pink, as I recall, with large black flowers, and it made up into two rather splendid, and when wet somewhat revealing, gaudy pairs of trunks which arrived in good time by post. Somehow we managed to escape detection by crowding in the centre until we had actually marched out, turned smartly right and lined up for our dives into the pool, and I will admit that, in the unlikely event of being allowed to continue, we did had an extended plan, in which Pat was going to fall in with much thrashing of arms in good Laurel and Hardy fashion before the command to dive, and I was going to break ranks and stand on the side watching and laughing and offering my assistance, but we never had to go into the water in the event, and were unceremoniously pulled away in full view of the audience, as a poor actor might be removed from the stage with a large hook on the end of a pole.

ENTERTAINING THE PUBLIC

And the gentle young maiden was barely sixteen
When we showed her the works of our threshing machine

When Parents' Day came round again, those of us from the senior entries without visitors were enlisted to demonstrate some of the displays and equipment that we toiled over weekly in the classrooms. A year previously I had spent the day in the coal yard with my yacht borrowing friend, so this was most definitely very much an upward career move as far as I was concerned. It was assumed by those in authority that we were reasonably technically competent by this time, and had a sufficiently well-developed sense of responsibility to put on a good show for the mummies and daddies. It goes without saying that they were quite mistaken on both accounts of course. When the selected group had been marched to the technical site, Pat was allocated the task of displaying the airfield flasher beacon, a massive piece of self-powered illuminating equipment mounted on a lorry chassis, which is used to flash in Morse code the airfield's identifying letters to aircraft above waiting to land. Pat had been chosen because, some time prior to this, on a most special publicity visit to our humble establishment by England's current entry to the Miss World competition, 'Ig' had been the chosen operator of the very same piece of equipment. As this young beauty, followed by her retinue of fawning R.A.F. admirers and press photographers, had appeared in our electrical training bays, 'Ig' had duly switched on this piece of apparatus exactly on cue, but the unexpected glaring, brightly flashing red neon lights, and the loud clatter of the contactor had thoroughly frightened the young lady and the press photographs, which sadly never made the front pages but were delivered to our barrack room never the less for our perusal afterwards, only revealed an obviously rather alarmed young female under the lecherous gaze of a wall to wall background of red illuminated, overly bright-eyed Fifty-Fifth electricians. For all the world it looked like the last moments before what today would be called a 'gang bang', and could hardly be regarded as good publicity. The blame for all this had been quite rightly laid firmly at 'Ig's' doorstep, and hence a new substitute performer was being given his moment on the stage.

Pat's idea of a demonstration proved to be rather bizarre. He did not even bother to switch the apparatus on, but instead climbed on to the highest part of the tower and began opening and closing his overcoat in rapid tempo as the visitors approached down the road. He obviously had in mind quite a different interpretation of flasher from his predecessor. 'Dadda' Brett, the electrical flight sergeant, was obliged to intervene and call Pat down from his high pulpit, and allocate him to something else less conspicuous.

My own speciality for the day was to be bomb gear, and I was to demonstrate all the equipment normally fitted to a Lancaster bomber, but which was now fixed to a huge wooden panel and fully wired up to bomb racks complete with dummy five hundred pounder bombs. After pointing out the various items of equipment and explaining their function, I invited questions from the audience. One interested parent asked me how it was possible to get rid of all the bombs in an emergency, and I demonstrated for him by pressing the jettison bar on the bomb distributor. Of course then all the bombs crashed to the floor. The watching crowd nervously scattered in all directions and I was left suddenly standing alone amidst the jettisoned bombs. The departing parents were treated to the sound of one of my assisting colleagues who began to curse me loudly in extremely coarse and vulgar language, since we would now have to lift and remount the bombs on the racks. That did seem to be too much trouble in view of the sudden lack of interest by our audience, so we wrote an 'out of order' notice on a piece of cardboard, hung it on the nearest bomb and departed after then to see how the others were doing.

'Horace' was showing guests that aircraft wiring systems, whilst complex, are quite harmless to operate because of the low battery type, twenty-four volts. Always at home when he had an audience, like a fairground barker, he was inviting guests, particularly older ladies, to switch things on and generally get their hands on the equipment and asking if anything was happening yet while he connected various parts of the circuit to the battery power behind the board. They were all thoroughly enjoying themselves assisting this most personable young man as far as I could see, and responding very positively to his questions from behind the board concerning what precisely was happening now. However, there was always a certain element of impatience in 'Horace's' character that could not resist temptation for too long. A kind of 'never give a sucker an even break' mentality as the Americans put it, and he soon clipped on a 250 volt insulation tester out of sight of his audience and gave the handle a brisk crank as he had done with poor Robin Berry a year or so earlier in classroom training. Elderly rheumatic fingers were snatched hurriedly away from the wires in no time at all and he lost their attention somewhat. We certainly enjoyed it but I cannot speak for all of the guests.

As we all began to grow rather rapidly in size, problems began to occur with our clothing. What had been snug and well-fitting two years before was often no longer quite so comfortable. 'Horace' arrived somewhat later than myself at an overcoat problem, which, with the collar fully buttoned to the top for ceremonial purposes, now pinched him until he was almost blue in the face and rendered normal movement almost Frankenstein's monster-like. Wiser than myself and with a great deal more cunning accumulated over the preceding two years, he now stubbornly stood his ground about buying a new one however, and finally, when he had been marched down to the stores under escort to do something about it, with good Scot's frugality he had even persuaded the camp tailor of all people to let him have a second-hand one in good condition at a reduced price. He came back from the stores that day, I remember, inordinately pleased with himself for some secret reason or other, and carefully folded this garment in its metal former and placed it in his locker with loving care. When autumn came around, out from the barrack block one chilly day stepped 'Horace' clad for the first time in this item of finery for all the world to see. That its previous owner had been a flight sergeant was quite clear because the three stripes and crown of that lofty rank were the sleeve badges now on display, which Horace of course had neglected to remove, and the former owner had also been a Canadian if the equally still intact shoulder patches were to be believed. In a manner quite befitting such a high rank, 'Horace' then began a serious inspection of the whole entry starting at the far left before our disciplinary sergeant arrived on the scene. It was a while before he even recognised who in fact it was out there gesticulating and shouting for people to stand to attention and indicating with pointed finger the badges which demanded such respect, and it took such threatening and persuasion to get 'Horace', who knew exactly how far to play this, back into the barrack room and to provide him with a pair of scissors located desperately in the C.O.'s desk drawer with which to reluctantly remove his newly found authority.

There had been altogether a rather significant change developing by the mid-point of our time at Halton, you see. Whereas in our earlier days the individual apprentice had been easily picked off in short skirmishes with authority, we had by now learned not only how to right some of the worst excesses of injustice by the odd spot of joint action, but even how to occasionally go over to the attack when we bothered to put our minds to it. We had all long learned to hone the small skills, such as verbal 'bar rattling', with which any visitor to a zoo is familiar, and by which any currently non-vigorous but never the less dangerous animal can be roused to its normal full fury by taking any convenient stick and dragging it violently across the bars of the cage. We were no longer single helpless sheep but extremely aware and astute, calculating young hooligans if need be. In all competitive sports and games, there is a point reached where the psychological advantage clearly passes to one side or the other, and which then leads on to victory, and we had already reached that point, although the enemy was seemingly not yet aware of the change that had taken place.

To my cynical amusement, I was myself enjoying a period of relative popularity with the wing personnel, a not-too-frequent happening. On a random inspection by our disciplinary sergeant Dick Corser one Saturday morning, he had selected me to be the one to receive the benefit of his extra care and attention yet again. I had never quite been one of his favourites since the day that he had discovered my early attempt to 'bash my bull', and he had had the opportunity to enjoy the pleasure of my company on many occasions when he was the duty N.C.O. and I had been on jankers. On this particular occasion he had removed my ceremonial belt and literally taken it to pieces, trying to find some small eyelet or inner area not polished and, having failed, had gone on to have my cap badge removed and the inner sides of the securing pin and eyelets similarly checked. I had not been too long off jankers by then and so he was absolutely wasting his time. He was so surprised and impressed that he awarded me a thirty-six hour pass there and then on the spot. I went to his squadron office later that morning and told him that, whilst I appreciated his generosity, I could not possibly get home and back on the pass, but could he perhaps instead ask the C.O. if I could accompany him on his next flying trip. There was no point in looking a gift horse in the mouth after all, and privileges of any kind rarely came my way. He agreed to do so, but a week later he took me on one side for a quiet moment and most regretfully told me that the C.O. had declined.

From that moment on, Dick's attitude to me changed completely, and soon after, largely due to his influence I fancy, I was selected to be the 'Stick Man' of the month. This was a supposed honour for the smartest apprentice in the whole wing, in which he was given an engraved ceremonial brass baton to polish and carry for the day, and, after a private interview with his Wing Commander, he had morning coffee and biscuits, non-chocolate variety I am sorry to say, with the Station Commander's wife in the chintzy blissful sanctity of the living room of their home, before a private chat with the Air Commodore himself in his office and the Head of Schools in attendance. All in all, it was a kind of short career review with small talk thrown in. Considering the amount of effort it took to get my appearance and the shine on that damned brass stick acceptable, I didn't see too much honour in it at the time. Much later I heard a corruption of the Stanley Holloway classic which describes it all so much better than I ever could:

Wi' a stick wi' an eagle on't andle
The best that the N.A.A.F.I. did sell

Additionally, the Saturday that I was chosen for this honour happened to coincide with a full kit inspection, which as usual required one's full efforts the previous evening as well as the normal domestic Friday night chores. It goes without saying that I was excused absolutely nothing, and my plate was therefore full to the point of overflowing, and necessitated my laying my kit out on the bed that night, and sleeping on the floor. In the actual event, it all went off rather well, particularly in the one-to-one chats, where there was some small possibility of exchanging viewpoints, although my honest appraisal of our early schools education hardly made a friend of the head of that department. Perhaps if all this had happened earlier in my first year, I might have had an easier apprenticeship, but it all came about in such a random manner that I still couldn't put such great store on it all.

A group of our entry had been selected for cookhouse fatigues one day under the supervision of their old friend 'Jim' who happened to be the Orderly Sergeant that day. They were in fact on the point of removing their overalls and leaving after completing the always miserable detail, when their ever-lustful young eyes focused for a moment on the latest addition to the cooking staff, a young auburn-haired Irish girl of all too obvious charms, as she entered the area. She swept disdainfully past them, head held high like a queen, and into the tin room where the cooking utensils that they had just been cleaning were stored. This was all duly noted by our good supervising friend 'Jim', who as usual was ever anxious to give them yet another lesson. He leered meaningfully at them before slowly following her into the room, and closing the door behind him. After a very short silence there was the most tremendous crash. The door flew open and she re-emerged with a very angry expression on a very red face and stormed away at a very high rate of knots. The still open door revealed 'Jim' flat on his back among the cooking utensils and no longer smiling at all. The news of this profound moral setback swept through our barrack rooms like wildfire and warmed many a young heart that night I can assure you. The glad tidings were that it appeared that, contrary to recent joint experience, there was a God after all, and although it had not yet been established that he was on our side, he apparently was most definitely not on theirs.

COME ON IN, THE WATER'S LOVELY

So roll on the Nelson, The Rodney, Renown
We can't say the Hood 'cos the...

Like those never-ending television sagas, summer camp time came round again. On this occasion we went further afield, to Little Marlow on the Thames. In later years I happily business-lunched within sight of what was, for me, a place of infinite horror at this earlier time. Some people are naturally tent people and others are simply not. The former are inclined to the boy scouts at an early age, and to things of nature, whilst others like myself, with a town streets' background, can derive no great satisfaction from being precipitately dumped in what they consider to be the wilderness. If it hasn't got a handy 'chippy' then it isn't even worth bothering about in my opinion. This time at least the weather was to be more kind to us, and did not interrupt our many comings and goings at all hours. Three events still stand out clearly in my mind.

The first was our long route march to the rifle range and it took many hours. We had stopped for our short regulation ten minute break in one village green, and 'Ig' used the opportunity to nip away and call at a cottage to ask if his water bottle could be filled, since against expressed orders he had already felt obliged to drink the lot. The friendly lady of the house had asked if he would like a cup of tea as well, whilst happily obliging him by filling his bottle, and he of course gratefully accepted, but he was sadly not destined to enjoy it. Ever watchful 'Jim' came along looking for him, chased poor 'Ig' away and then stayed on to drink the tea himself. On our eventual arrival at the range, a special treat was in store for us. We were to be allowed to fire the bren machine gun, it seemed. What a joy! In the fullness of time I found myself in a prone position behind a bren and with twenty others or so stretched in a long line on either side of me. Of course one did not simply lie down and start blasting away. One waited for the words of command as always.

"One magazine at your target in front, range two hundred yards, in short service bursts. Wait for it! Wait for it!"

It was then that the totally unexpected occurred. From the right-hand side of the range, a group of squealing pigs suddenly emerged and began to cross ahead of us. I am sad to say that the long hours of training in waiting for the word of command then proved to be quite valueless, as the whole line of us opened up on the unfortunate animals as one. The noise and excitement prevented our hearing the shouts to cease fire, and absolutely no one stopped before his magazine was empty. When the great silence finally descended, the pigs exited stage left, not a single one so much as scratched by a volume of fire at least as great as that used on D-Day. The regiment people didn't know where to begin in their tirade against us, either for firing without orders or plain bad marksmanship, but, since it was their responsibility to make sure that the range was safe from dangerous intrusions, it was reluctantly decided to call it a draw. The lorries which had been promised to pick us up after the long march there did not of course materialise. A by then boorish familiar ploy and quite expected, so we started our merry way homeward with the customary encouraging calls to, "Sing or Double!" After such long practice, I now actually preferred the double myself. It gave fine opportunities to observe the growing discomfort of the regiment persons and so to anticipate that piquant moment when they would be the ones to have to call a halt.

The second incident was the last battle in my ongoing war with that certain regiment sergeant. I may unwittingly have given the impression that his three years of punitive effort were exclusively spent on me. Nothing could be further from the truth, however. He simply lashed out at anyone with an apprentice's wheel on his arm who chanced to come near him from the very first day. My friend was giving instruction this particular day in the art of crossing rivers in rubber dinghies. We had been entertained yet again by his selection and persecution of some fractionally less attentive creature than the rest of us as the butt of his questions, and the, by now, boring final advice, "Don't larf at eem, feel sorry for eem ees stupeed."

We had all long ago come to our own conclusions concerning exactly who it was that was 'stupeed'. Finally he was ready to demonstrate the ease with which this water crossing could be accomplished, and it certainly could be if you just sat there while another poor creature did all the paddling. You may guess who was selected as canoe man of the month. As I dragged the wretched rubber boat to the bank of the Thames after having to inflate it by hand, he walked alongside, all polished boots and gaiters, making coarse observations and remarks all the way, and thoroughly enjoying himself, but such was his concern with his fruity performance that his attention was not focused on me particularly so he failed to notice my loosening of the boat's rubber drain plug. My companions in this class were far more alert to such possibilities, and settled back with obvious relish to watch the emerging fiasco. He got in, I got in, we both got in together as the song goes, and the two of us set off for the far bank, with my responding to his invective, abuse and orders with as great a demonstration of wild arm threshing and splashing as has ever been witnessed on Blackpool beach in high summer. This display of dynamism greatly pleased him, although the course was necessarily rather erratic. We were somewhere past the mid-point of that great river which has witnessed so much history when it started to become increasingly clear that perhaps all was not well in the nautical sense, and he bade me loudly and anxiously to return whence we had just come. My classmates on the shore then came in on cue and took up their own role, with wild shouts of encouragement and raucous ribald advice and laughter and slowly, to their and my own great joy, we sank. It was worth the cold ducking and the immersion in evil-smelling mud just to see him crawl up the river bank. I, as a gentleman of course, offered my assisting arm which he angrily declined, but in doing so only he slipped again in the mud. By my score I made it an emphatic 4-2 but he abandoned play from there on in. There was newer, softer, less revenge-seeking meat for him to handle in the junior entries.

The third incident occurred when, near the end of our stay, we were visited by no less a personage than an Air Vice Marshal, who wanted to see us in action for himself. He was obviously the very keen type of leader who liked to be at the very heart of whatever happened to be going on, such as a public execution or something of that sort. He had made a wise choice as it happened, because it turned out to be what I would describe as a typical 'Horace' classic. We had been well briefed for the visit and a demonstration of a two section infantry attack was to be on display. One section of ten was to pin down the enemy position frontally with fire from blanks, whilst the other section did a smart right hook attack. This involved the crossing of a low fence which was topped by barbed wire, but the height was such that it was easy to vault over using only one hand and holding your rifle in the other. With an eye to avoiding unexpected developments no doubt, I had not been chosen for either the attacking or defending groups by our good friend the regiment sergeant. For once I remained with the rest of my entry, in the at ease position behind the Air Marshal and his group of supporting officers, watching the display. Our sergeant friend must have had an unusual lapse of memory or, perhaps more likely, was by now reduced to employing people with what he considered to be only marginally tainted records, because 'Horace' of all people was one of those selected to be in the right hook attacking group. After the frontal attack group started to bang off their blanks of covering fire, the others rose to their feet as one and set off to dash the final few yards and vault the fence. 'Horace' proceeded to the fence all right with the others and at the same frenetic pace, but then seemed to be overcome with a sudden attack of conspicuous doubt as he reached the wired obstacle. He began to laboriously climb it as an old age pensioner might cross a busy road at a pedestrian crossing, and only then after selecting one of the support posts as a suitable crossing point with rather undue care. He managed to reach the top long after his section colleagues had disappeared into the shrubbery beyond, but his trousers had somehow now managed to get caught at the very top and his efforts to unhook them began to make him sway alarmingly. The Wallace family jewels were plainly in grave danger.

We all realised by now, that this was another of 'Horace's' instant farce situations, but our friend the sergeant plainly did not. His little gem of a demonstration was about to come apart at the seams unless he acted quickly to recover the situation that this buffoon was creating. He lost his temper and, purple-faced, ran over to where poor 'Horace' was still oscillating. All eyes focused on the little act within an act drama which was unfolding. With one swift kick of his ammunition boots, he attempted to dispatch our poor friend 'Horace' to the other side of the fence. There was a tearing sound, plainly discernible to us observers, then an agonising cry of, "Oh me goolies!" hence the rather late change to an alternative nickname, and our friend fell to the other side of the fence. After a moment he reappeared in sight clutching his nether regions and wailing, plainly no longer interested in any right, left or centre hook attack. Everything after this was an anti-climax. Our main difficulty was to not burst out laughing in front of the great man, particularly when he was overheard to remark that it had all been very illuminating.

It was this sheer inability to detect when they were being set up, and their absolute and total belief in 'by the book', that made the regiment people so vulnerable. I only remember one occasion when our friend the sergeant showed anything approaching a sense of humour, and even then it was not intended. He was insisting that anything at all could be brought to perfection by means of repetitive drills. Of course some youthful impish voice from the safe obscurity of the back row enquired, "Even sex?"

To which our friend replied in the affirmative. "Of course. Eets just whip it een, pause, whip it out, pause, and wipe eet."

We had not been back from summer camp for very long when a very big organisational change took place and one with far-reaching consequences for us all. The whole of the apprentices were to be divided up by trades with the airframe fitters in number two wing, the engine fitters in one wing and the ancillary trades, electricians, instrument-makers and armourers, in three wing. This was the final act in the attempt to crush entry spirit. Fearing it rather than trying to encourage it, and with no ability whatsoever to direct it into useful channels, this was a typical response and achieved nothing of the sort. What it did succeed in doing was to make at best for a very uneasy truce in each of the wings. Although all the Fifty-Fifth electricians except the N.C.O. apprentices were now reunited in one room again, there was the question of the rights of senior entries, now in multiple. It had long been the unquestioned privilege of senior entries to go to the front of meal queues for example, but now this implied that the whole queue would be in strict entry number order and there were just bound to be problems. It wasn't long before my friend Pat Cropley went to the front one lunch time, and a Fifty-Sixth electrician foolishly tried to stop him. Pat could be very violent as I have said, and the other fellow went down with a head butt that required stitches whilst Pat, unmoved, stepped over the recumbent body and took his plate. There was open hostility and it certainty went on for the rest of our time there. Ken Smith was the 'snag' in charge of the Fifty-Sixth room at that time, so you can guess how difficult that particular incident made life for him. Some entries were tougher than others and held their ground well, the Fifty-Sixth electricians in particular being always a bit of a handful for us, while with the airframe fitters it was apparently the Fifty-Seventh who provided the main opposition.

Despite the lack of involvement for most of us in any educational certificates any longer, we had to try and make some sort of decent showing in our final schools examinations or suffer the consequences in our final overall result. As I have described, many of us were in a woefully bad position, but at this time we had a stroke of real luck and got a new teacher, Flying Officer and later and most deservedly Flight Lieutenant, 'Mickey' Gilbert, who really did try to put right what had been neglected over the previous two years. He gave up his free time in the evenings to coach and prepare us for the finals, and every single one of us owed him a great deal. He also happened to be a talented footballer who was on Gillingham's books and was a member of the Halton station team which reached and won the cup final of the R.A.F. station teams that year.

Another player was 'Paddy' Howley from the Fifty-Fourth entry, and himself on Tottenham Hotspur's books. I once saw a very frustrated 'Paddy' take a football into an empty barrack room adjoining our own and smash every window with consecutive shots from alternate feet. All the station personnel were allowed to go on the coach trip to see the final, including the apprentices, and many of us did so to support 'Paddy' and P.O. Gilbert. We stopped at a pub on the way but the apprentices were noticeably in a separate group from the airmen. We couldn't have afforded to buy a pint between us even if we had been allowed to, and there were plenty of the 'they' about to make quite sure that we didn't. We did not waste our day entirely though, and during the match, when the ball was kicked into the spectator area which we occupied, it rapidly disappeared under the nearest greatcoat after an even more rapid puncturing. We may have been poor but we were resourceful, and that ball was a rare prize for us to use later.

RISKY ENTERPRISES AND MORAL WELFARE

There was an old monk of great renown

The Flight Sergeant electrician in charge of our workshops training was Flight Sergeant Brett and he was a really good chap, always genuinely concerned with the welfare of the electricians. He was by no means soft, but he was always scrupulously fair. He was accordingly affectionately known to us all as 'Dadda'. He occupied the N.C.O.'s room outside ours and although he was a technician and not of the disciplinary people, he kept a fatherly eye on what he regarded as his boys. He was a bachelor in the tradition of many of the old-time N.C.O.s and his evenings were mainly spent in the sergeants' mess propping up the bar, which tended to make his morning ablutions noisy affairs and often matters of some urgency too. He enjoyed the privacy of a separate toilet for his exclusive use, which unfortunately also made him rather vulnerable to our attentions. One of our number removed his toilet roll one evening and replaced it with a single sheet of extremely rough cardboard. The following morning, well before reveille, we were awakened by a loud moaning shout from the ablution area. We all snuggled happily in our blankets listening and awaiting the trumpet call that morning. The next evening again the exchange of rough cardboard for soft paper was made and the following morning the cry was even louder and more distinct: "Oh no! The little bastards have stolen my bog roll again!" No one went to offer assistance or condolence but we did decide to prudently quit while we were still ahead.

Our barrack block at that time was situated right at the edge of the wing area, very close to the site of the old tin N.A.A.F.I. which was still in bleak disrepair, and the closest point to the R.A.F. Hospital, where every Sunday night a dance was held. We were not permitted to attend this, it goes without saying, but some of our room used to remove their apprentices' badges and cap bands, and slip out to spend a couple of hours with any W.R.A.F.s or nurses that they had somehow managed to meet and charm. It was also the custom to slip into the doorway of our block on the way back from the dance for a quick spot of slap and tickle with the young ladies. One Sunday evening after the departure through the back window of two such romantics and well after lights out, 'Dadda' paid an unexpected torchlight visit to our room, no doubt making sure that all of his children were safely tucked in for the night. Two beds were empty and he patiently waited in the shadow of the open window for their owners to return. The rest of us, feigning sleep, also waited in delicious anticipation. We heard their arrival from the whispered giggling of the girls as they moved to the block doorway for you know what, and it was then that a thoroughly alarmed 'Dadda' leaned out of the window and began to shout, "Large! Evans! Come in at once! Get away from my boys you old bags!"

The girls ran off in terror and that source of romance was nipped in the bud for the rest of our time there. Although 'Dai' and Tony both received a good shouting at, and dire warnings concerning their obvious lack of morals, 'Dadda' did not charge them or put them on punishment, which was quite untypical of the other N.C.O.s. Now painfully aware that young libidos were apparently beginning to try and run rampant, 'Dadda' kept an even wider eye out for potential trouble of this kind, and he soon found it too. As a corporal apprentice, 'Ig' Noble had the privilege of a room of his own, which also afforded him a certain amount of extra scope in popping in and out at late hours unnoticed which was not available to the majority of us. With a long term view to the general good of his metal health, he was keeping himself warm with a young nurse in the barrack block doorway one evening when ever vigilant 'Dadda' came down the road from another session in the sergeants' mess bar, and spotted them. The young lady immediately fled and, his exit path blocked by the advancing 'Dadda', 'Ig' was obliged to beat a hasty retreat through the back door of our block and across into the entrance of the adjoining one with 'Dadda' now in high pursuit. Up the stairs galloped 'Ig' and he burst into Ken Smith's room who was in charge of the Fifty-Sixth electricians and lying in bed at the time. With no time for an explanation, 'Ig' opened the window, got his leg across the sill and climbed down the drainpipe to safety. Since 'Dadda' was hardly the climbing type any longer, he only got as far as the window to watch his quarry escaping. He knew full well who it was but again nothing further happened about it. 'Ig' was singularly pleased with himself after this, and even took to a certain measure of boasting about the exploit so a group of us visited him late one evening when he was already in his pyjamas in his own room, tied him to his bed and carried the whole thing down the stairs and deposited it in the road outside. The implicit message being, 'So get out of that one Houdini'.

We had progressed in our trade education to the stage of learning about the wonders of modern automobile electrics at workshops, and many ancient and decrepit vehicles, obviously recovered far too late from bomb sites, stood on rusting rims awaiting our tender attentions when we had finally acquired some basic knowledge of the components and how they worked. Well, that was sort of happening in the background, but the large physical area required for these many vehicles on which we were supposed to be carrying out our practical work offered splendid opportunities for some horseplay too, undetected by the instructors. Robin Berry, for example, was detained by 'Ig' for most of one afternoon in a camouflaged Hillman car, the wiring of which had been ingeniously modified to apply the spark from the ignition system to anyone attempting to open the door and get out, and how poor Robin tried. That sort of modification would make a fortune now in burglar proofing cars I realise, so perhaps I should revise my opinion and we were perhaps inadvertently learning something after all.

Pat Cropley had made a truly wonderful discovery. We had been provided with a heavy metal-wheeled wooden trolley with which to drag our piled tool boxes from classroom to classroom as the subject of our technical education periodically changed. If a daring driver was prepared to lie full length on this vehicle, steering by means of the long iron-pulling bar held out in front of him like a battering ram, and relying on the propelling power of ten classmates pushing as hard as they could, then some very hairy drives between the cars was possible. Pat became the virtuoso Stirling Moss of this vehicle, and could almost skim the paintwork from car after car as he hurtled round the track at breakneck speed. I joined him for some time in a daring circus duo act, which included my death-defying leap from the roof of one of the stationary cars on to the trolley as it flashed past. The act continued in fact until the day that I tore one trouser leg almost off on a projecting nail in the trolley. My bare white leg swinging away amongst all the blue clad ones made in itself a good source of humour when we marched up the long road back to the wing that evening, but it was obviously too expensive to continue with that part of the show. It was all very thrilling stuff, however, when the instructor had nipped out for a quick cigarette and a break.

Pat's daring display had not passed entirely unnoticed among some of the junior entries in training in neighbouring bays, who began to eagerly anticipate their turn at initiation into the mysteries of automobile electrics and an opportunity to try out their own driving skills. When the Fifty-Sixth came to that particular part in their training, they lost no time in trying it out for themselves, but their driver apparently lacked Pat's deep sense of anticipation as well as high quality steering skills, and in his early headlong flight around the parked cars was soon caught out and barely in control at high speed. He was obliged to seek the safety of any convenient emergency exit road that presented itself and it was for this reason that he made a sudden unplanned skidding turn from the safety of his normal track into the corridor separating it from the next bay. It was pure bad fortune that poor 'Dadda' happened to be walking the other way at the same time, and he could only have had the barest of glimpses before he was mown down by the onrushing vehicle, the driver of which was quite helpless since the vehicle had no brakes. Poor 'Dadda' was lifted unconscious on to the trolley and pulled all the way to the sick quarters, as a dead monarch might be taken to his final resting place on a gun carriage. He returned with a broken ankle some weeks later but no one, including himself, was sure what had really happened so no action was ever taken.

I must have been at least a partially attentive pupil on the part of our training which attempted to teach us something about airfield lighting systems and flarepaths, yet another part of the electrician's area of responsibility, wouldn't you know, because the technical information as well as the initiative training that I had been daily undergoing until then was later to stand me in very good stead indeed. The occasion was nearing the end of my service, when I had been requested to provide some sort of colourful flashing disco lighting for the forthcoming sergeants' mess ball. I was a trifle more artistically ambitious than that, and in addition to the simple lighting began to envisage something much more spectacular, like a centre piece of an illuminated water fountain. Accordingly I had used my position as i/c ground electrical section to closely examine the wiring of the runway lighting system, and to temporarily detach and 'borrow' one of the sodium light units without bothering to disturb the flying control officer, within whose jurisdiction these sorts of things lay, with such a trifling detail. Being a ground to air missile station, absolutely no one, friend or foe, ever flew anywhere near us day or night out of simple prudence in my experience. I also called in a favour from my friend the air electrical section sergeant boss, and obtained the loan of an aircraft fuel pump and battery. My last call was to the safety equipment section which had a rather pretty young W.R.A.F. corporal in charge, and, after a certain amount of practising my rather rusty chatting-up skills, I persuaded her to loan me a rubber dinghy in return for a personal invitation to be my guest at the ball. We had lots of young, single and handsome senior N.C.O.s living in at the time so I didn't anticipate enjoying the privilege of personally entertaining her for very long, and she could then make her own selection and proceed from there as she felt inclined. It did however take considerably longer to explain all this much later to my wife and to convince her that my intentions, whilst perhaps not quite falling within the category of being totally honourable, were most certainly not romantic. Armed with my odd gathered items, I set up the water-proofed sodium lamp in the water-filled dinghy in the mess entrance hall, surrounded it with plants and flowers which helped to conceal the battery, and set up the fuel pump to provide a cascading shimmering display through my garden watering can spout over the bright yellow light. It was indeed most spectacular and promised to make the whole affair infinitely more memorable when I switched it on at the soon-to-follow ball. What I did not know at that time was that the air traffic control officer had somehow noticed, God knows how, that he had a flare path unit missing, and that he had approached my own young ground engineering commanding officer with his problem about what to do next. Two days before the mess ball, my young leader called me in enlisting my aid, and asked if I could possibly track down the evil culprit, find out what had happened to the missing lamp and possibly recover or somehow replace it, or his friend, on whose inventory it was recorded, was likely to be firmly in the unmentionable. I was able to assure him that it would take a man such as myself no more than three days at most to satisfy both their needs, and save their careers.

The night of the ball, I was stood in the lounge doorway dressed in my best uniform finery, sipping a welcoming drink in the company of my attractive safety equipment guest, the envy of all who passed by, except my wife of course, and the both of us admiring the spectacle of what we had jointly provided, when I noticed the arrival in the foyer of none other than my young C.O., who some idiot must have invited quite unknown to me. Even as he was removing his coat, I realised that it obviously could be only a matter of minutes before he would set eyes on my masterpiece; indeed, he couldn't possibly help but notice it, and I envisaged some rather deep understanding growing quickly within him after that. Rising immediately to the emergency, I advised my charming guest, "For God's sake get over there and pull that silly bugger or we are both right in it." Fortunately she was an astute young woman and did as I bade her without too much further explanation.

The following morning my young noble leader sent for me, and I entered his office waiting for the axe to fall, but no. He leaned his chair backwards and smiled at me benignly.

"I take it that it must have been you, but thank you so much for the kind invitation, sergeant," he began. "I suppose that you must have noticed the young lady that I was with most of the evening. Well, the truth is that I can safely say that she is obviously rather fond of me."

"So are we all, sir, so are we all," I added hopefully, beginning to suspect that I might after all just be safe.

"Actually I called you in to ask a personal favour of you, sergeant," he continued, "I hope that I can rely on your discretion to keep all this quiet, because frankly I intend to go on seeing her, other ranks or not."

I was fortunately able to find it within me to so reassure him, and as I departed, he added from the heights of his euphoria, "Oh, by the way, just get the lamp and stuff back as soon as possible, will you?"

A single phone call was sufficient to reassure me that my pretty young friend had not after all been required to make any kind of noble sacrifice for both of us, but, however, she apparently did not wish to receive any further invitations to the sergeants' mess social events, and I was to return the dinghy immediately, "Before the sickening ****** is round here after me again."

HOW TO TREAT A LADY

She was poor
But she was honest

Anything outside the monotonous daily routine of apprentices' life could be absolutely relied upon to either involve extra sweat, extra discomfort or both. One notable exception to this was the occasion that we were visited by a very senior W.R.A.F. officer, doubtless feeling the width of the young material being produced for future consumption by her young ladies out on the stations in the big real Air Force. We were all assembled in the station cinema for an address by this most formidable lady, who was of the iron-haired forty-going-on-eighty type, and supported not so much on legs but rather by a massive undercarriage that would have delighted the designers of the four engine bomber. As she was being assisted on to the stage by a young officer, our friend the regiment sergeant moved down the shadows of the aisle with a deep frown, making surreptitious 'psst' noises and indicating by his fluttering hand movements that we should encourage the poor old dear by a little polite applause. It was all very confusing; after all, he had been spending considerable time and effort since the day that we had signed on in eliminating any such gentlemanly tendencies and responses, so it was a very thin and rather uncertain applause that began to ripple around the cinema. I believe that it was Tony Large who made the day, albeit with perhaps a slight touch of over-reaction. He began to applaud loudly with much enthusiasm, as one might for a modern pop artist, and he was soon joined by his fellow electricians who were not slow to realise that yet again an opportunity seemed to be presenting itself. The applause became louder as understanding began to dawn row by row, and there were soon a few whistles to add to its strength. The poor old dear positively beamed with delight. It probably was the absolute high spot in a very long and dull administrative career, and she even chanced a little curtsey on those massive legs in the centre of the stage. Perhaps Tony went a trifle overboard with his clearly discernible chant of,

"Get em off! Get em off!" which of course produced some laughter amongst the more vulgar minded amongst us, but the sudden general enthusiasm was quickly taken up and began to make itself more than heard.

The sergeant, however, was not in the least misled by these events. He was by now fully cognisant of blossoming anarchy when he saw or rather heard it, and he was amongst us in a trice trying to detect exactly who was the main instigator of all this, most suddenly instantly restored to his normal state of mind.

"Right, yew little bastards, I'm warning yew. I know what yor game is and I'll soon fix yew," he muttered grimly as he forced his way down the rows.

We all beamed at this fine compliment. In any case it was already too late, as both we and he were well aware. The fat lady was already on the stage and the opera could not now end, nor obviously could punishment begin, until she had sung her song. I do not recall the particular subject of her discourse to us that day so it probably was important, and most certainly had nothing pertaining to sex in it or I certainly should have remembered. Actually, when it was all over, it was my humble opinion that we had done her a great favour. Like the aged spinster who took in a male lodger, she had obviously enjoyed every moment of it, and it couldn't have been a very frequent occurrence in her life.

That evening after lights out, from the comfort of our beds, we mused over the events of the day, re-savouring the magic moments, and discussing the strange sudden blooming of character on the part of the sergeant but Tony, ever prematurely wise in the ways of the world, summed it up for all of us before settling his head down on the pillow for sleep.

"Perhaps it is just that he fancies old scrubbers," he explained.

PUTTING THEORY INTO PRACTICE

The moral of the story, it is this
Always have a shuftee before you have...

Our technical education was entering better and more interesting phases all the time. There soon came the first of two visits to the airfield for a period of instruction on real aircraft electrics. This was intended to be the practical application of many hours of instruction in classrooms on the intricacies of aircraft wiring systems such as power circuits, bomb gear, and the like, or so we anticipated. What actually happened was that for openers we again sat in a classroom, this time tented, learning by heart the numbers of R.A.F. forms and how they were to be completed. Eventually, however, like all things, even that passed, and the day came when we were turned loose on some of His Majesty's flying equipment at long last. Can you imagine it – real aircraft. We couldn't wait to get out there, I remember. Not to examine the wiring or carry out any servicing or any boring stuff like that of course, but to get in the pilot's seat and handle the controls or swing the gun turrets or look through the bombsight. So it was that, when my class were told that our task for the day was to carry out an inspection on the four engine Lancaster bomber stood outside, there was a mad rush. Not as fleet of foot as some of my fellows on this occasion, I found all the turrets occupied and swinging wildly in every direction when I finally got on board. The pilot and engineer positions were also full with at least two in each, and every imaginable switch and lever was being pressed and controls tested for movement. Even the wireless operator's and navigator's positions were in full use with queues in waiting. The only remaining vacant place was down in the nose in the bomb aimer's position and Paddy Swoffer was already on his hands and knees moving into that just ahead of me. When he had wriggled in, his hands jerked nervously around him, seeking for something, anything, to manipulate. His anxious gaze fell on a metal ring in the floor and he grabbed it and pulled it in one swift motion. Unfortunately this was the release for the emergency exit on which Paddy himself was standing, and before my eyes the floor fell out with him on it like some Aladdin on a faulty magic carpet. There is a considerable height from the nose of a Lancaster to the concrete below, and Paddy was the next one to suffer a broken bone.

The Fifty-Second entry, attending their second spell of airfield training just after our first visit, were obviously more disciplined than we were, and their engine fitters had actually been attempting to do something useful by starting up the engines of a more modern Lincoln bomber. This was most unfortunate because something went badly wrong and it caught fire and, probably due to the still less than skilful attentions of many previous generations of apprentice electricians, ours included, the fire extinguisher systems failed to work. The whole thing burned to the ground where it stood. That also made the local newspapers, complete with pictures.

I found myself beside a plaster-casted but most unrepentant Paddy Swoffer in a twin engine Mosquito one day. He felt quite confident that he could start the engines completely unaided, although our scheduled task was to test the navigation lights. As Paddy put it to me, "You have to learn to see the big picture, Grim."

The propellers began to turn and the engines fired briefly, but the starter motor on one engine jammed, and, no matter what he tried, the propeller continued to turn, accompanied by the shriek of the starter motor gears being stripped to bare metal. We both hurriedly abandoned the aircraft in a manner befitting the very best of aircrew emergency drills, pulled out the plug of the ground electrical supply as we went, and departed elsewhere at high speed in search of something else to play with.

I can unfortunately only recall learning one useful fact during our time at the airfield, and heard of just one other which actually occurred in our improvership year at St Athan, but was never the less useful to bear in mind. I will deal with the later first. The occasion was when, as on all flying units, the morning and afternoon tea breaks occur as the N.A.A.F.I. van arrives rather than at any specific hour and one must keep an ear tuned for the traditional cry of "NAAFI Up!" if you happen to fancy a wad and a char, that is.

One of our number arrived late in time to join the queue at the van, and saw that the one in front of him was bent down apparently tying a shoelace. Being impatient and not realising that anyone dressed in overalls might be in the queue, not just the ex-brat mates that he had become used to over the previous three years, with the typical vulgar familiarity in use between us, he extended his index finger and made a swift jab at what he hoped was the centre of the bent down fundamental orifice in order to hasten the queue, whilst simultaneously shouting out, "How's that for centre?"

The young airwoman straightened up rather smartly, exceedingly red-faced, turned and smacked his surprised face, and departed for her tea break elsewhere. The rest of his comrades gave him a generous round of applause as a consolation prize.

The other occasion was when a group of us electricians were miserably attempting to install some wiring in a tented portable hangar in a group practical exercise which was intended more as a punishment than a lesson, since we appeared to be intent on destroying the aircraft when released on them. Someone inquired where on earth the toilets were located in such a primitive piece of equipment, to which one of our number replied by saying, "Here" stabbing a hole in the canvas with his screwdriver at an appropriate height, extending it to a slit with one tearing pull, unbuttoning his flies, pushing his item of personal equipment out through it and beginning to enjoy a good long pee, with the expression of happy release on his face appropriate to such an action. Unfortunately he had chosen a rather inappropriate moment to demonstrate his version of the famous Brussels fountain because there was the sound of what was most obviously an officer's voice calling from outside and saying indignantly in the very best Queen's English, "I say! Who on earth is that in there?"

I can only imagine the impact that the sight of this unexpected sudden event may have had on delicate sensibilities when viewed from outside. The item of personal equipment was hurriedly snatched back inside and someone asked in a whisper who it was outside. I personally declined to put my eye to the slit to find out, for hygienic reasons alone, and we all jointly decided to prudently take ourselves elsewhere and with some urgency.

Link Trainer training was supposed to equip us to be able to service the flight simulators of the time, which again fell within an electrician's area of responsibility in those days. Our civilian instructor, a Mr Hogg, was as equally concerned with keeping a good polish on the floor as with technical matters, and he kept this very much in mind when any of us chanced to irritate him in some way, usually with some wild and unauthorised testing of his trainers. One day it happened to be 'Swing' Swoffer practising high G stall turns and spins, and when he had been removed from the cockpit, the glossy floor was pointed at and the instructions went as follows, "I want to see it shine! Ai, shine! Like a cat's arse in't moonlight," Mr. Hogg informed the culprit in a broad Yorkshire accent.

These simulators were not quite as yet the technological electronic wonders of the present day, but still technically rather demanding, however. I eventually learned everything that I knew on this particular subject from on-the-job training when I was employed in the synthetic trainer section during the one year improvership at St Athan which followed my apprenticeship, but at this early time, the temptation to play with these new toys rather than learn about them was simply too great for all of us. Link Trainers had been built to accommodate only one, but I found myself in the pilot's seat with at least three others on several occasions.

This in fact proved to be excellent training for a later time, when I had, for one of my many sins, been detailed to take care of a trainer at an exhibition in Cardiff during my improvership year. On that occasion I was playing the part of the skilful, handsome and daring young aviator, standing up in the cockpit and flying with one hand quite nonchalantly, and all that sort of nonsense. A rather attractive young lady came up and expressed a wish to try and fly the trainer, with my able assistance of course, and so, as we both squeezed in, I pulled the blind flying hood down just to make it really interesting. My good friend Pat Cropley chanced to be assisting me at the exhibition, and noticed too late for once that I had scooped him. He also correctly concluded what was going on inside the cockpit under the blind flying hood by the somewhat erratic and wild flying manoeuvres of the machine. Jealously he began to beat on the side panels with his clenched fist, but to absolutely no avail. I kept the hood firmly shut and reassured the young lady that it was only the simulation of hail and bad weather, and we clung even closer to each other for safety. It was a rather happy and lipstick-besmirched 'Grimbo' who eventually finally emerged, to find his friend Pat with hands on hips on the bottom step watching enviously.

This was also the day that some spectator asked if there was a charge for allowing his wretched child to sit in the machine. Pat, quick as a flash, answered, "No, but of course benevolence is always appreciated". It was just loud enough for everyone in the audience to hear and with just the right tone of sincerity. No one enquired precisely which benevolence he was referring to, and we made over five quid each that day, and on all the other days that the show went on. Charity begins at home, after all. However I am digressing from our final days at Halton.

As part of our final year rewards, we were dressed in our best uniforms and taken by a splendid, highly polished R.A.F. coach to have a look around a manufacturer's works for the day. My class went to Plessey at Ilford, who made a great many electrical components for aircraft. To keep an eye on us, we were accompanied by the officer in charge of electrical training, a Flight Lieutenant Pitts, who was soon to become much more familiar with the whole of the Fifty-Fifth entry not just the electricians, and one of his esteemed sergeant instructors affectionately known to us as 'Barrow' Yarrow, from his East End of London origins. The morning was spent looking round their assembly benches, and of course chatting up the young girl assembly line workers when opportunities presented themselves, which they most certainly did. We had all learned much from our early days of queuing behind our elder brothers, the Fiftieth, in the 'tank,' and just how to flash the teeth and to stand upwind to allow the females to catch the scent of our aftershave. The afternoon was intended for a closer look at the marvels of their research laboratories and the wonders of their test facilities.

As lunch time drew near, the factory manager arrived, drew our officer on one side and enquired if he would care to lunch with him in the management dining room, which it seemed that he would, and they departed. The works foreman then arrived and asked the sergeant if he would like to eat with him, in the works canteen, and they walked off chatting happily together. No one asked us anything, so we went to the nearby pub, frequented and recommended by the girls in the morning conversations. We promptly jettisoned the already tired-looking sandwiches supplied to us for our midday repast, courtesy of His Majesty, and began instead to enjoy the drinks being bought for us by our new-found and more prosperous wage-earning friends. The whole day suddenly began to take on a rather more positive and sunny tone. I was dancing closely with a charming young lady of recent acquaintance, who asked me coyly why some of my comrades had stripes on their arm whilst I did not. Fearing that she might be planning to gravitate to what she suspected was upmarket material, I had to quickly improvise. My explanation that they were wound stripes shocked her, and she went on to ask if they had been shot or something. I replied that it was common knowledge among the rest of us that every single one of those so chevroned had something missing, but I went on to reassure her that I at least was alright on that score.

When the suitably fed officer managed to finally tear himself away from his convivial host and eventually found time to seek his sergeant, they soon jointly discovered that we, their joint charges, obviously must be elsewhere, and in the fullness of time, when we did not materialise, they were finally obliged to go in search of the still missing subordinates. Eventually, a credit to their homing instincts as well as to the abysmal lack of good pubs in the Ilford area, their humble efforts were finally blessed with success. The party was in full swing when they walked side by side through the swing doors of the saloon bar. Some of the girls were already ensconced on best blue knees, and others were wearing our orange banded ceremonial caps at jaunty angles as they danced with the jacket-unbuttoned owners, myself included. It was all very friendly with no sense of impropriety whatsoever, you understand.

Alas, we never got to see the splendid research laboratories nor did we manage to cast as much as a glance of the eye on the wonderful test facilities. Indeed, Plessey by afternoon was to remain an unfathomable entity to the Fifty-Fifth electricians, because we were all put back on the coach and returned at once to Halton, accompanied in the front seat by a silent and plainly rather irritable officer, and immediately behind him and slightly to the rear of course, in accordance with good order and service discipline, by a somewhat miffed sergeant who had doubtless been planning an evening at home with good old Mum, Dad and the family fruit and veg vehicle. Our request that we be allowed to sing to brighten up the return journey was rather churlishly refused, I recall. As the sergeant so eloquently put it, "This isn't any bloody day trip to Blackpool." There were supposed to be two outside trips like this for each entry I remember, but for some unaccountable reason we never did get the second one.

There was a certain unfortunate sequel to our quickly terminated Plessey visit. We returned to Halton too late for normal tea time, and still hungry from happily abandoning the sandwich repast prepared for us that day in preference to a more liquid diet. We ate a solitary makeshift meal which included two seven pound tins of jam and bread for dessert. We thoughtfully decided to retain the remainder of one tin for future enjoyment, and this was entrusted to 'Ig' to keep in his room where it was less likely to be discovered. We were all then free to steal bread from the cookhouse at will as always, and freely pop in on occasion for a spoonful or so just to make things more tasty. Unhappily the tin was discovered in 'Ig's' room under his bed the very next afternoon on a random casual inspection by 'Tug' Wilson who was seeking yet again to ingratiate himself with his Commanding Officer, and our friend 'Ig' then soon found himself under the escort of his fellow corporal apprentice 'Eddie' LeGrove, and standing before the latest and last of our Commanding Officers, New Zealand-originating Squadron Leader Dempsey, more familiarly known amongst us as 'Kiwi' or 'Black Jack', depending on your personal standing with him at the time. 'Ig' was charged with stealing a tin of jam, the property of His Majesty. The vital clue linking 'Ig' with the whole wretched affair, apart from the fact that it was found in his room, was that it was his spoon, with his serial number on it, that lay handle up in the tin when it was discovered. It became obvious at once that our leader was not overly interested in pursuing the matter further if he could possibly help it. Under questioning, the corporal cook called as a witness could not positively identify this particular tin as the one issued the previous evening.

"Why did you have the jam in your room then, corporal apprentice?" asked 'Jack' after finally ascertaining from 'Ig' himself the source of the forbidden fruit, which was most obviously not stolen since it had been issued to us as part of a daily ration.

"To put on the bread, sir" replied 'Ig'.

"Bread! What bread? There is nothing on the charge sheet here about any bread. Where did you obtain this bread and for what purpose?" continued 'Jack'.

"I was hungry, sir," replied 'Ig.'

"Dat's right, sor," interrupted 'Tug' in his best Irish, still trying to obtain a conviction. "Der little buggers is always hungry."

At this point, 'Jack' became far more serious.

"This is a most disgraceful and extremely serious business, if apprentices are obliged to obtain food in this manner," he went on. "And I shall have to look into it further. You are admonished."

'Jack' had found a very convenient way of abandoning the whole wretched business. Perhaps they were themselves tiring of tormenting us because the punishment was almost non-existent on this occasion, although that was the absolute last that any of us ever saw of His Majesty's jam, however.

'Kiwi' was a good deal more understanding than some of his predecessors in yet another case. 'Charlie' Rushforth's girlfriend, after a long illness, finally died of cancer whilst we were taking our final schools examinations, but with 'Kiwi' and 'Mickey' Gilbert behind him, he was sent straight off on leave. We all realised that both 'Ig' and 'Horace' had needed outside intervention in almost identical earlier circumstances, which perhaps accounts for my nickname change.

Life still held unexpected and unjust surprises for us however. We electricians had been told that we would be taking an examination paper on a certain Monday, leaving the whole intervening weekend for some revision. Came Friday, and to our surprise we were marched to the schools hall for the examination and sat down in our places. Then the inevitable arguments and protests started in serious earnest, and were interrupted by the arrival of 'Pussy' Funnel who had just had his leading apprentice's stripe taken from him for some trifling minor infraction or other. That was an unfortunate catalyst, since 'Pussy' was rather well liked, which further inflamed the situation, and finally we all then jointly refused to take the paper. The sergeant in charge was completely out of his depth, and left hurriedly for Workshops seeking assistance, leaving us under the eye of a certain corporal ex-boy entrant who kindly and privately advised us that we might be fractionally less exposed if we at least wrote our number and name on the paper with a number one in brackets for the first question. Still prudent if rebellious, we did just that and sat back. 'They' arrived in streams shouting, threatening, pleading and warning. Only one of us gave in and started to answer the questions on the paper however and finally, due to our intransigency, the test was suspended until the promised following Monday. There was a certain degree of natural justice finally, in that, as we correctly anticipated, over the weekend there was simply no time to prepare a new paper and the very same one which had been set before us on Friday was presented to us again on Monday with much higher than average results than had been expected except for one of us of course, but he was rewarded for his loyalty by a further promotion in the N.C.O. apprentice hierarchy.

HERE COMES THE AIR VICE MARSHAL

Stand by your beds men
Here comes the Air Vice Marshal

Now that we were nearing the end of our apprenticeship, we began to dream of that magic day when some Air Vice Marshal or other would come and take the salute at our passing out parade and we could leave this accursed place at last. He would probably be called Sir Scrotum Cummodley or something similar, would have a large bucolic red nose from years of overgenerous consumption of the port on mess dining-in nights and most certainly would have the D.S.O. and bar and D.F.C. and bar and had never caught anything from a toilet seat in the whole of his distinguished service. We had long become familiar with the kind of officer chosen for these kind of ceremonials but we didn't much care any more. We just wanted him to come soon. There was a long tradition that after the final school examinations the whole entry marched back to the wing area via the bays of all the workshops. The pipe band always led this march and, although it interrupted ongoing lessons for a brief moment, it also gave a splendid boost to the morale of the junior entries under training in the workshop classrooms, reminding them that one day it would be their turn too.

Our march back did not begin auspiciously, when we were forbidden to pin our examination number cards on our tunics in the customary tradition, courtesy of the 'no rock cake without milk' deputy head of schools, still waiting and longing to be the real head by the way. On the actual march back the regiment N.C.O.s tried to prevent our turn into the workshop area for the march through. It was like peeing into the wind. File after file was ordered to carry straight on. File after file ignored the order and was sent to the rear. It was a wilful point-blank refusal to obey an order by two hundred apprentices. After marching through workshops the entry as a whole had also decided that they would march back to their original and symbolic home, number three wing, and not to their by trades wings. Again this is exactly what they did. It must have been a sobering thought to our masters that entry spirit, despite their best efforts, was still apparently not quite dead. I am quite sure that action was contemplated because the defiance was so blatant, but by this time not one of us could have cared less.

As not quite but almost fully grown men, it was considered by our masters to be perhaps an appropriate time to take us to the quiet intimacy of the drill hut at Henderson parade ground, the service equivalent of behind the cycle sheds, and give us a lecture on sexual matters in general and venereal diseases in particular delivered by the Medical Officer. Such was our devoted level of theoretical interest in the subject at the time that the vast majority of us could undoubtedly have marked his card and given him a few interesting pointers as well. He kept it simple to begin with, starting with the birds and bees, moved on to the rabbits and so forth, and there was a short film at the end showing the happy smile but unfortunately diseased organ of some poor American airman. When the lights came on and the final and inevitable questioning period arrived, 'Swing' Swoffer, ever anxious to demonstrate that he at least was not like the rest of us, with our simple primitive lustful interests, but was however keeping himself fully abreast of all scientifically related factors, asked if it was true that the disease could in fact be caught from monkeys. The M.O. regarded him long with growing suspicion, plainly not a little anxious since this was after all an apprentice doing the asking, before enquiring gravely, "Tell me, what is it exactly that you are thinking of doing?" We all thoroughly enjoyed that.

Not everything was funny, however. Tony Large, the best electrician, had become so disillusioned in his final year that finally he refused to complete a workshops examination paper one day, including the number and name part. He was court-martialled and got twenty eight days in a military prison at Colchester for his trouble. He could then have been barely eighteen years old. The system could be very harsh, particularly when dealing with the individual and wanting to make an example of him. Tony was fortunate not to be put back an entry for a month's break in training at that time, which would only have further prolonged his misery but on his return he confided that in the main it was much the same as that we had been undergoing daily for the past years. In earlier days I understand that apprentices were even publicly caned, so maybe he got off rather lightly. Eventually, in his improvership year, I believe that Tony bought himself out of the service when allocated to the repair of domestic electrical equipment in a final attempt to further humiliate him, and again it was all very much the R.A.F.'s loss because, as I have said, he was the very best of us.

There must have been a few extra quid in the Air Force budget that last year of ours at Halton, because the R.A.F. was going through a period of refurbishing all the apprentices' cookhouses, from now on to be known as dining halls. There were polished tables, table mats, glass water jugs and even hanging basket floral arrangements. It was all most civilised and completely unexpected, but there were a few inevitable drawbacks of course such as an apprentice from each room being detailed, as yet another extra task, to lay the table prior to each meal, and the odd hiccup in the cooking staff system too. One could, in theory at least, now help oneself to soup from a central tureen and similarly to gravy for example, but it was noticed rather early that this was in fact the self-same tureen labelled soup on one side and gravy on the other. Bill Mercer, one of our 'plumbers', was quietly enjoying his soup one day when one of the overhanging flowers chanced to drop into his bowl. The prissy uselessness of it all so angered him that he instantly reverted to his former philistine self and lashed out at the offending overhanging basket with his knife, and in a trice totally destroyed the armourers' floral display for which offence he was put on a charge and awarded a few more days' jankers to add to what had already been a rather distinguished criminal career. His C.O. remarked at the time that apprentices' dining halls must be exceedingly dangerous places.

My new-found hand fitting skills came into urgent use just weeks before we finally passed out from Halton. I was joining in a room game of tennis ball football one day with the door as the goalposts when I received a marvellous pass from 'Slash' Gwilliam. One of those that you simply cannot resist, and I let go with a gorgeous rising left footer which unfortunately just scraped the top of the crossbar, but hit the tannoy loudspeaker mounted above it full on. There was a godalmighty crash and the wooden cabinet lay in small splintered fragments on the floor with the badly dented speaker among it. This looked like big money plus certain punishment for me at a time that I needed it the least. However, using the only materials that I had available – cardboard and insulating tape plus some paper clips borrowed from the C.O.'s office – I managed to get the whole thing reassembled, albeit in a very shaky fashion, and back in its perch. Of course it no longer worked, but by leaving the door open one was able to hear at least some of the messages that played in the room across the way from our own. The very real danger was that the whole precarious thing would crash on to the head of the first one who banged the door, and then all would certainly be discovered, but my comrades were extremely cautious when entering or departing and sort of slid through with raised eyes keeping a watch out for any sudden developments. It became somewhat of a race between the insulating tape glue retaining its stickiness and our departure. The glue won, but I wonder who did eventually get the whole lot on his head.

YOU HAVE GOT A GREAT FUTURE AHEAD OF YOU, YOUNG MAN

Oh the eagles they fly high in Mobile
Yes the eagles they fly high in Mobile

The final half year was a very serious business indeed to me and most of the other electricians. I still sometimes have unpleasant dreams in which it plainly figures. I studied every evening well before the official revision period began, and my only serious concern was the fitting exercise which would be set for us. In the mock trade tests carried out by our normal instructors, I knew that I was doing rather well when one civilian instructor, unable to catch me out in even the smallest of matters, with great irritation finally asked me, "Well what is the reference number of the grease that you would use then?"

I didn't even blink an eye, but answered with the volume number and part of the correct technical manual where I would find the necessary information. My overall assessment, even by my former oppressors, was that I was rated as a likely A.C.1.

The whole of our final workshops examinations were carried out by a party from the Central Trade Testing Board at R.A.F. Chigwell and it took well over a week. They were mainly warrant officers with the odd flight sergeant or so thrown in, and were all electricians of long experience and under the command of a young flying officer as the senior board member. Being extremely thorough in doing my homework, I had watched from my classroom seat the earlier trade tests of the Fifty-Third and Fifty-Fourth entries going on around me in the workshop bays with great interest, and, knowing most of them very well by this time, had taken the opportunity to seek them out afterwards, enquire and make notes about what the questions had been and directed by whom, whilst it was still fresh in their minds. In this way I had a very good idea by the time it came to be our turn for this modern inquisition exactly who were the toughest examiners as well as the usual likely questions. I had also noticed that the board officer occasionally carried out an oral test himself in borderline cases, and to check out the marking of his team. 'Pansy' Potter of the Fifty-Fourth entry informed me that this had happened to him and that the officer in question had rather a dry sense of humour and apparently a great in-depth interest in the subject of batteries. I had stored the details of this piece of whimsy in my mind, suspecting that if the choice was alphabetical then it was probably going to be either Robin Berry or myself for this extra session. Whilst some of us were taken away to be questioned individually on a particular subject, the rest would be working on the fitting exercise. The tests were extremely fair and not subject to interference from the instructors, so our friend Mr Tripp's opinions were fortunately not an issue. The fitting exercise selected for us was to make a steel bearing extractor, case hardened tip and all. It involved a fair amount of lathe turning as well as hand tool work, and proved in practice to be a difficult one to complete in the permitted time scale, particularly for the less able like myself, so we worked evening overtime hours to complete it. The more speedy and skilful of our number, after completing their own test piece, kept it out of sight and worked on making the more difficult parts for the others, quite unknown to the supervisors. Tony Large certainly did more than a fair amount of my lathe turning for example. I was also selected, as I had half anticipated, to enjoy the added privilege of having part of my oral tests repeated when the board officer carried out his own independent checks on the marking being awarded by his own board examiners. Then I learned about the dry humour.

"Tell me apprentice," he began as we walked down the workshops together, "what is your favourite subject?"

"Power supplies, sir," I, truthfully, answered.

"Good show," he said admiringly, since this was well known to be quite a difficult subject. "But I prefer batteries and that is what I intend to ask you about."

We then soon got down to the composition of the very last ion in every molecule of sulphuric acid but, thanks to my tip from 'Pansy', I had prepared myself well and rode sublimely smoothly through it all.

Finally it was all over and soon after we were all assembled together at schools, as an entry, to hear the results. The overall result determined your future rank and pay, and was made up mainly from your workshop result, partly from your schools result and, we feared, partly from your wing report too. I was more than anxious, because it was obvious that my workshop result would have to make up a lot of leeway for the other two, and I knew it. Pat Cropley was another in exactly the same position. A really exceptional electrical apprentice, perhaps one in every three entries on average, could be made a Leading Aircraftsman but even then only after six months of further service. A more usual top grade would be Aircraftsman First Class for perhaps the very best top three per cent. The rest would be Aircraftsman Second Class. The results announcements started from the best, regardless of trade, and worked their way down, so you can imagine the tension, waiting for your name to be announced to see if it was within the higher groups. It was perhaps the only time that I experienced the whole entry in a mass being reduced to total silence. I was one of six of the electricians who made AC 1, 'Pash' Page failed and went down to the Fifty-Sixth, and the rest were AC 2s.

The best of the results supposedly were soon being given preliminary interviews to determine which, if any, might be selected to go forward for a full commission interview. There were a few others included too, plainly based on quite other criteria. With my track record, I knew this to be a complete pure waste of time in my case and probably for all of our top six electricians too, with the exception of 'Ig' perhaps, who was a corporal apprentice. In the event 'Ig', 'Curly' and Robin from the electricians were eventually sent forward to Ramridge House for final selection, and 'Ig' was the only one of the three who had made A.C.1. Finally there were just two technical commissions granted in the whole entry, to 'Ginger' Robertson and Andy Kidear. It was somewhat later in my career that I myself was eventually sent forward for a selection board, but I believe that what transpired then was not so different. This was not quite the same as the induction interview for Halton which had gone so well for me, and where they had actually wanted you to join. Now they were all too obviously intent on finding reasons to reject you. I put on my serious responsible face for the occasion, since smiling is well known to be a sign of early lunacy. After the usual preliminaries of having to wait to be asked to sit down, trying to overhear their quiet furtive whispering together, and some general paper shuffling, all of which were intended to induce a degree of nervousness, the three officer interview board on the other side of the table began.

They liked to get you in a corner as quickly as possible and watch you wriggle.

"The situation is this, ah, ah". Slow reference to first paper. "Oh yes just so, Bristow," the elderly Squadron Leader president began, then leaned forward confidentially. "Not by any chance related to Warrant Officer Bristow, I suppose?" he asked.

"Yes, sir. My father's younger brother," I lied at once, having never heard of such a person. The truth was that the only other Bristow I had ever even heard of serving in the whole Air Force was 'Wee' Bristow, a Sixty-First entry rook who had been in Robin Berry's room and therefore someone even lower on the ladder than myself, but it always pays to have connections in high places, I had learned, and a Warrant Officer is after all the highest rank in that other Air Force.

"Well, you are the captain of a bomber crew on an extremely important mission, Bristow. Your best friend is your navigator and he is seriously wounded on the way to the target. He will undoubtedly die if you do not turn back. What are your actions?" he asked, his pitch rising at the latter part of the question. I was smart enough to spot the general trend that there was no correct answer, just more and more cornering, and I hedged my bets by playing delaying tactics whilst I had time to think out where any answer might be about to unfortunately lead me in the next couple of questions. Privately I already did not fancy any of our chances of getting back at all from this mission, the way things seemed to be going.

"I would strap on his parachute and throw him out, sir, and trust that the enemy would act within the spirit of the Geneva Convention. Then I would proceed to the target." The president was not to be so easily diverted from his nasty little scenario, and nor were his assistants who took up the action.

"Quite impossible, I am afraid, Bristow. You are flying over shark-infested waters," interjected one.

"In any case, the enemy is of the despicable type who would not hesitate for a moment to torture your friend in order to find out your intended target," went on the other.

"Please continue," added the third.

The despicable ones would have to be a bit swift off the mark too, I thought, if they were going to fish my old pal out of the jaws of some great white in the nick of time, realise who he was and get the old thumb screws on him before the rest of us ran out of fuel, but who was I to argue?

"I would therefore regretfully have to continue to the target, sir," I went on, anticipating the coming attack regarding my apparent lack of moral values and rather cavalier approach to the general well-being of my friends and crew. I added nobly, "After all, it is quite likely that we will all be killed, and that is an acceptable risk that all aircrew must be prepared to take."

I could easily afford to say so because there had, as yet, been no call for aircrew volunteers from my entry so no one who mattered was involved. I would have been home and dry with this line too no matter what they threw at me, since all officers have a deep love of nobility without possessing a shred of it themselves, but as usual I made my fatal error by going on for one more sentence: "After all sir, none of us can live forever."

This piece of homespun wisdom seemed to strike a previously unsuspected and extremely unpleasant chord deep in the old president's intellect, if such a thing exists among officers, and he seemed to take it that I was endeavouring to make some kind of oblique humorous reference to his advanced age. Up to that moment, he had plainly been planning to go on at least until the end of the world interviewing candidates such as myself, and my remark irritated and at the same time somehow depressed him. One could almost see his lips moving, "Can't go on living forever," in sheer disbelief. He suddenly became introspective, even morose and took no further interest whatsoever in my interview, no doubt preparing himself for an interview of his own with a much higher authority. It seemed for a brief moment that I had somehow managed to shoot down one of my opponents quite accidentally en route to this hypothetical target, and I began to feel more pleased with myself, even reassured by this unexpected turn in events. Assistant number one, he of the shark-infested waters, and who wore no aircrew insignia, also chose to take my piece of simple philosophy amiss, and interpreted it as an attempt at a subtle put-down reference to the fact that he had plainly survived the war sitting behind a desk. These were all very touchy, hypersensitive people, it seemed, each with strange, barely concealed hang-ups of their very own, and I came to the conclusion that perhaps one needs to be at least slightly mentally deranged in order to obtain a commission. In their eyes I now became not merely a hopeful candidate but an extremely slippery and thoroughly unpleasant young villain, noble actions or not. Both 'shark' and his colleague had noticed their leader sitting between them mentally sucking his thumb whilst he went down in flames, and they renewed the attack with all guns blazing. They made verbal slashing attacks both singly and in co-ordination, making every possible attempt to persuade me into all sorts of nasty extremes as the situation that they laid out for me became increasingly more horrendous. The operating height became lower and lower as the aircraft suffered more and more damage, and the answers required suggested remedial actions such as throwing the whole crew out, unfortunately without the parachute option, because I had discarded them earlier in an attempt to gain height. At last through the flak, night fighters and searchlights I began to catch a glimmer of the ultimate question and the suggested response that all that was left for me to do was crash the aircraft into the target with myself still on board. I wasn't going to be so easily trapped that way, however. If nothing else, my prior three years at Halton had taught me that to wilfully damage even the smallest piece of equipment, which was, after all, the property of His Majesty or his heirs and successors, was a very serious matter indeed, and would most certainly land you in deep trouble. So they did not quite manage to corner me and I did not quite manage to get a recommendation from them, which I considered to be an honourable draw and the best that could be hoped for in the situation.

In the afternoon, there was also a practical side to the selection process which consisted of joining teams of five strangers all dressed in numbered overalls and being required in turn to be the leader attempting to navigate the group through an obstacle course. My turn came with us all stood on one concrete block with some poles and rope, and being required to move the whole team to another concrete block without touching the ground. Although mechanical engineering as a whole had never quite been my forte, this fortunately did not involve distinguishing between a pull and a push, and under my able direction we managed to get one man out on the end of a pole, with the rest of us acting as a counter balance, to erect a pair of sheer legs between the blocks, complete with rope and to get him back. Finally I gave the order for the first man to swing Tarzan-like across to the other block. Unfortunately, the rope had been fixed just a shade too long and he crashed headlong into the side of the far block and slid gracefully to the ground, out cold. In a flash, I gave the order for number two to go, but he seemed to be overcome with a sudden malaise of reluctance similar to that suffered by poor 'Horace' at our last summer camp, and so, on my first real attempt at leadership in action, like the captain of His Majesty's good ship *Bounty*, I encountered a small mutiny. I was not so foolish as to attempt the next crossing myself, of course. I was being interviewed for a commission after all and was expected to show at least some officer-like behaviour. Suffice it to say that to this day I have received no word of how well I did on a points scale of one to ten. Like the draw for the next round of the F.A. Cup, there is no report of how well the losers played in the last round.

THE BALL AT KURRIEMUIR

You could nah hear the bagpipes
For the sound of swishing...

Another joint parade was held under the auspices of the regiment sergeant to obtain volunteers for first aircrew and second overseas posting. Those willing were supposed to step forward one pace for each option at the word of command as I understood it but there was some sort of misunderstanding because, whilst I stood perfectly still, reflecting on how on earth he intended to distinguish between those who only wanted to be aircrew and those who only wanted to go overseas, almost everyone else, very sensibly not wishing to volunteer for anything at all, took the elementary precaution of stepping at least one step backwards, which only goes to show what can happen to you if you are inclined to think too much.

One of our entry and an electrician too, managed to achieve for himself a remarkable position in the hall of all time apprentices' fame by being the first ever to be rejected from the aircrew volunteer tests for cheating. As he explained it to me later, the major part of the pilot selection process consisted of sitting in a row of individual cubicles, each of which had a cathode ray tube display and a control column which was to be used to control a spot of light forever moving about the screen under some external random process, the objective being to keep the spot within a surrounding square of light. This was apparently a relatively easy task, but there was an added distracting element in that a separate red lamp which came on randomly had to be changed as quickly as possible to green by the manipulation of a switch. My friend was astute enough to notice that if he moved his chair a shade backwards he was able to observe that the red light was being applied to each cubicle in strictly cyclical order, and that in such a position it was a relatively easy matter to anticipate its arrival at your own cubicle and to have your hand ready at the switch accordingly. What he did not appreciate was that the score was being maintained live, and it was not long before seasoned examiners were abandoning their posts in order to watch this new successor to Mickey Martin, Leonard Cheshire and Guy Gibson in action, at which point all was most unfortunately revealed, and my friend's prospective aircrew career was abruptly terminated.

When everyone had eventually returned from their interviews, there began the rehearsals for our passing out parade. On one of the earliest of many inspections, it was noticed that, due to my somewhat tainted criminal past, I wore no good conduct stripes on my best blue jacket sleeve at all, whilst most of the others had at least two and some even three. This matter was corrected by hurriedly issuing me with chevrons to be sewn on that very night, so that I would not be required to tell my sad tale or offer any explanation to any overly curious inspecting officer, and I became an honest man at last, at least in appearance. For my last and precious few apprentice days, it seemed, I was to be indistinguishable from the purest of the pure. We were given severe warnings about our conduct on the great day.

The Fifty-Fourth had made national newspaper columns when, at their passing out parade, they had fired blanks from their rifles over the parade ground. No one had explained to the watching crowds that the rounds were only blanks, and this had perhaps understandably caused some alarm amongst both the more genteel and the decidedly guilty. If they had used live rounds and aimed at the crowd then I might perhaps have taken them more seriously as an entry, but they always were reluctant to go for the throat when the chips went down in my experience. Nevertheless, the rifles were examined and the culprits were locked in the guardroom on their arrival at their new camp. The Forty-Seventh had burned their entry number into the tarmac of the parade ground with lighted dope when they departed, and the Fiftieth had done their best and almost succeeded in electrocuting the orderly sergeant as I have described, so perhaps the track record indicated that some kind of trouble was quite likely.

Nothing special was anticipated from us because a new ploy was being used to prevent any more regrettable incidents from occurring. We were to be the first entry not to proceed to our new camp directly from leave. Instead we were to return to Halton, and then all of us would be moved together. The implicit threat in this will not be lost on you, I am sure, as it was not on us. However the very best of planning can be brought to naught if only you have the will and the time to think about it, and our particular farewell to Halton was accordingly planned to take place actually on the joint move, when the system was likely to be at its most vulnerable.

Although our entry was now separated by trade and located in different wings, we decided that we would still have a traditional Christmas as our last one in the apprentices, and we followed the general format taught us earlier by the Fiftieth. The presents were presented by Roy Ellis our instrument-maker comrade and our selection to be the master of ceremonies. We were all already aware that he had failed to pass out and was to be put back to the Fifty-Sixth entry, but we wanted to make the point that regardless of pass or fail, in our eyes he was still, and would remain, very such one of us. The Wing Warrant Officer was now 'Shorty' Parkes, most certainly not even an embryo 'Steve' and he was given a pair of stilts. Our electrical instructor friend Sergeant Yarrow got his very own barrow and was therefore now able to contemplate a future of having his very own business separate from Mum and Dad. The barely literate Flight Sergeant 'Red Eye' Reid got a comic which he must have just about completed by the time that I finally left the service eleven years later, and 'Black Jack' got a pair of garden shears since we could not locate, let alone steal, the sheep-shearing variety intended. It still felt to me that we were now very much a divided group rather than a total entry, and accordingly it left much less of an impression on me than any earlier service Christmas. Something much more relevant was obviously at hand and it filled my thoughts.

On the passing out parade day, mummies and daddies came from as far away as Scotland. Ernie Baldwin and other earlier deportees from our ranks rejoined their old entry for the day to a rapturous welcome. It all went off like clockwork. It's quite a feeling to march off through the massed ranks of the junior entries all at the present arms I can tell you, to the sound of the pipes playing 'Auld Lang Syne' and to realise that it is all for you. Three years is a long hard climb for everyone, even the best-behaved like myself, and it's rather nice to be the centre of admiration and attention even for just one day. Of course my father and stepmother didn't come, but that was no big surprise for me. In any case I wanted to be where I felt that I belonged – with my friends, and to share this special day of ours with them. There was the prize-giving after the parade, both the official and the unofficial. Tony Large deservedly got the prize for best electrician, and I got the one for most punished electrician. It was a close-run thing between Pat and myself in the end. He had endured more formal detention in cells, but I had a longer consecutive jankers' period which, finally just tipped the balance in my favour. The banners which we had secretly made in the evenings came out of their secret hiding places at the side of the parade ground, and were proudly displayed for all to see on our march back to the wing. Colourful as they were, parents and guests gathered close, full of curiosity to see them as they fluttered proudly in the wind. 'Slash' Gwilliam had spent many evenings of his very real artistic talent working on ours, which was made from a stolen bed sheet. The top scroll read '55th Electricians' and it had a French motto resplendent beneath its central heraldic symbol of a lightning flash ('in fosse' I believe is the correct term) splitting two testicles. It read 'Ascendez Votre Conduit,' which freely translated means, 'Right Up Your Pipe.' Such a fitting final mass opinion, I have always thought. The afternoon was left free to watch sports with the proud parents, and if you didn't have your own with you, then you joined someone else's.

The evening was for now legitimate boozing in the Wendover village pubs, which until now had point-blank refused to serve us, and was followed by the Passing Out Ball. For many unaccustomed to strong drink, it was just that. The romantics brought their latest conquests and the rest of us had to improvise yet again. Previously well-concealed and most unsuspected relationships with instructors' daughters and even local schoolgirls came to light, and the very well-built young lady whom we had all ogled on our way to workshops every day for the whole of our three years came on the arm of one of our airframe fitters, 'Blom the Bomb' Blomfield.

'Slash', who had been transported to the ball by necessity in a wheelbarrow found outside a pub, and still more than a little under the weather, was seen to be swaying on the fringe of a group of our officers and their ladies, silently listening to their small talk and beaming happily about him at everyone. Some opportunist wag called out, "Give us a real rasper, Slash," and 'Slash' lifted his leg high in true Chaplin fashion and happily obliged, as he had unfailingly done when requested over the three previous years. The group sniffed the air in all too obvious embarrassment, and rapidly dispersed.

'Horace', with typical Scottish gallantry, asked permission to request the pleasure from the Wing Commander's lady, and they began what he later chose to describe as a jitterbug. Personally I thought it looked more like he was trying to break her arm in a final act of malice and revenge. The poor woman, of course, was hardly in a position to decline, even had she known of the true nature of 'Horace's' skills on the dance floor, but he showed not a trace of pity as he grasped her hand and dragged her mercilessly round the dance floor, finally ending the display with his own individual war cry of a Woody Woodpecker cackle, which, though familiar to all of us, only served to further alarm an already very anxious perspiring lady. The rest of us didn't really care too much about dancing. I did not indulge with many of my colleagues in the traditional bed tipping of the junior entries after we returned to our quarters. It had always savoured a little too much of bullying in my opinion, and it was surely high time to put that sort of thing firmly behind us, and in any case I could barely stand up unaided, let alone tip beds. I believe that gentle little Robin virtually demolished the junior entry room of which he was the 'snag' single-handed, watched in silent open-mouthed wonder by 'Horace' and some of the others.

The next day we went on leave, and, what with the exertions of the many rehearsals, the parade itself, and a general lack of sleep, I arrived home rather fatigued and still ever so slightly the worse for drink from the ongoing celebrations on the train with my fellow townsmen from junior entries, who had seen fit to drink to my very good health and future excellent fortune. As a matter of fact, I was still vomiting as I came down Suggitts Lane where my father lived. I negotiated rather than opened the front door, and went straight to bed after easing my already aching head gently round the sitting room door for some very brief "Hellos", accompanied by a sudden violent attack of the hiccups. My father reserved his lack of approval, and the opinion that the Air Force had obviously failed miserably in educating me thoroughly, until lunch time the following day when, I am glad to say, I had recovered somewhat and was better able to bear it. I could not in all honesty help but feel that he was entirely correct after what I had just experienced for the past three years. Perhaps it awakened in him a certain sense of déja vu and visions of 'Brandy', his father and my grandfather, all over again. As an able seaman on the clipper Cutty Sark, 'Brandy' had apparently not been overly reluctant to celebrate his return from any long and difficult voyage either.

In spite of the fact that I was now nineteen years old, had no longer really been at home for the previous three years, and was completely independent apart from still being quite unable to darn my own socks, my father had still apparently made few adjustments or concessions to this situation, I soon discovered. I was still in some ways the velvet-trousered small boy that the next door neighbour had in her mind as far as he was concerned. I returned home one evening on that leave at barely half past eleven o'clock to find him still sitting up waiting for me.

"Where do you think you have you been until this time then?" he asked, tapping the clock. Thinking that it was perhaps some kind of guessing game joke or other, I replied with a gay light-hearted laugh that I had only been doing what he had probably been doing when he was nineteen years old. It was, it seemed, an unfortunately wrong answer because he almost went berserk on the spot.

"You what! Disgraceful!" he shouted, "We will talk about this again tomorrow!" and he stomped off to his bed.

By the morrow he had perhaps at least begun to realise that I was indeed growing up because nothing more was said about the matter, and from then on I was given the front door key when I was going out, via my stepmother's hand of course.

Nothing else was very special about that last leave as an apprentice that I can recall other than a feeling of immense relief that it was all over, and soon I was back at Halton waiting for the move to R.A.F. St Athan in South Wales. We all walked straight into the unexpected. We had of course all removed our apprentice badges, squadron discs and hat bands the day that we had passed out even if we had reluctantly been obliged to put them back on to go on leave. After our return from leave, our view was that we were after all certainly airmen now. The disciplinary N.C.O., it seemed, wanted them all back on, and we refused. The Squadron C.O. came out to us, and tried to persuade us to put them back on. We refused again, and then they just gave up. I suppose that it was inevitable that there should be one last confrontation, so typical of what had gone on daily over those three previous years. The next day we handed in all those extra items of kit that had occupied so much of our time, but I was cautious enough to retain the highly polished brasses just in case it started all over again. My rifle and bayonet were returned to the armoury to await a new owner. The next day was the last time that we would all parade together and it passed virtually unnoticed. Our kit was stacked on the parade ground and then loaded on to lorries and taken to Wendover railway station to be put on to the troop train which was to take us to our new home at R.A.F. St Athan in South Wales. We marched to the station together as an entry for the last time.

TAKE THE 'A' TRAIN

And the whole bloody issue was driven by steam

Now is my last story in the apprentice saga. We had made very good preparations for our journey. From an evening perusal of the contents of the squadron sergeant's desk after the usual weekly lock-picking, we had learned that we were to be accompanied on our journey by the Flight Lieutenant who had been the officer in charge of electrical workshops, and had enjoyed our day or rather half day at Plessey's together, good old 'Cec'. He was being posted to St Athan as a section C.O. To assist him, two disciplinary corporals were also being sent. Some of our entry had been detailed to ensure that the officer's kit was put on the train, and they faithfully did so. Of course it didn't happen to be our train, but they most certainly made sure that it did get put on; in fact some of it went on one train and some on another going in the opposite direction to make quite sure. Although I was not one of those so detailed, I can personally vouch for the fact, since I overheard at least some of the phone calls from the anxious owner, that it was some weeks before British Rail finally located it all and delivered it to its rightful place. It did not seem to help them very much that 'Cec' was unable to say on which particular train he had lost it.

With all of us from all trades finally aboard, the train departed, but was obliged to halt a few stations up the line when a signal box reported that there appeared to be a number of people climbing along the outside of the train as it passed his box. The engine fitters had been on their way, by this perhaps unusual route, to offer their fraternal greetings and assistance to the engine driver and his mate on the locomotive driving platform. It apparently became quite crowded just before the train was stopped.

Our officer was unaware of the reason for our delay because he occupied the only first class carriage on the whole train in splendid isolation and, anyway, the airframe fitters had taken the precaution of obtaining a railway carriage door key by somewhat dubious means, with which they locked the door after he had got on and was happily ensconced. This was preceded immediately prior to his arrival by disabling both the window sash and the corridor access door. If he chanced to have a call of nature during the trip then his choice of remedial actions was most certainly going to be severely limited. The two corporal assistants had begun to raise a certain amount of complaint about our general demeanour, so it had been left to the armourers under Bill Mercer, always ready for an emergency, to cool down their anxiety by re-accommodating them in the toilet and locking them in. The engine fitters, their willing assistance on the footplate declined, were persuaded to relinquish the engine platform and returned to their compartments. The electricians, who had thoughtfully brought along a selection of stolen tools, began an inspection of the wiring as soon as the train started again, and badly needed modifications were soon well under way when it was discovered just how badly British Rail maintains its rolling stock. Festoons of cables hung out like Christmas streamers from conduits for the length of several corridors. The armourers, anticipating the advent of terrorism by some decades, focused their attentions on the likelihood of there being a bomb concealed on board, which required a certain amount of panel dismantling, and the airframe fitters smoked in such volume that it was not possible to see anything through the windows of their compartments either in or out, a situation referred to as a 'good rest'.

The train, for reasons that I was never able to fully comprehend, was scheduled to stop at Swindon and Newport, although we were its sole passengers and had been forbidden to alight under any circumstances. Accordingly at each of these places as we pulled into the platform Roger Barratt alighted first playing 'The Hop Scotch Polka,' a popular tune of the time on his pipes, and the carriage doors then flew open at this clarion call and we all got off.

'Cec' was able to observe this wilful disobedience from the comfort of his first class accommodation, and became thoroughly alarmed. Unable to get out of his compartment or to open his window for some unaccountable reason, he obviously began to feel that he was beginning to lose control of the situation, and, standing at the window, began to wave his arms in a rather frantic manner in an attempt to attract attention. Those of us who were privileged to see him waved back on their way to the refreshment room.

There, as senior entry, we exercised our undoubted privilege to move to the head of all the queues, and of course there were a number of surly civilian objectors, just as there had been among the rook entries in the wing. A number of small scuffles and shouting matches consequently developed, and the speed of service at the counters consequently declined. The hungry ones of us further back then had to improvise their refreshments in the time scale permitted for the train's stay, and here 'Horace' yet again rose to the fore and set a positive example for everyone by snatching a sandwich from the plate of a very surprised lady and cramming it into his mouth, after raising his hat of course and smiling all the time. We were dealing with the civilian population after all and there was no need for anything impolite. An instrument-maker, taking up the action, added his customary two lumps to a gentleman's cup of tea and began to slowly and deliberately stir it. In no time at all it was as if the lunatic asylum was having a day trip out. The red-faced and sweating station staff under the direction of the station master, who, unfortunately somehow lost his gold-trimmed peaked cap in the process, finally managed to evict the last of us in order to get the train on its way again. Of course inevitably some of our number just didn't quite make it and had to catch a later train. The poor corporals could raise no assistance at all, despite their hammering on the toilet walls and must have had a most uncomfortable journey.

We finally arrived in South Wales later that night, minus the few stragglers of course. The pipers under 'Jock' La Haye formed up at our head. We hoisted our kit bags on to our shoulders, and we chose to march into Number 32 Maintenance Unit St Athan to the sounds of what else but our very own tune, 'The Bear'. No creeping into our new careers via the quiet back door for us, even if all the senior entries from the Fifty-Third down were still in residence there. Just everyone was being informed that we, no longer the senior entry but yet again 'cheeky bloody rooks', had at last arrived. Poor 'Cec' was in some disarray, and most unfortunately was too busy to accompany us just at that moment, having had to arouse the attention of the platform staff by beating with his hands on the compartment window until they had finally managed to get the door open, only to discover that he and his kit were now somehow parted. The corporals, I believe, did not manage to get out until they arrived at the terminating station two stops up the line. We had managed to have our day finally, so those junior entries still back at Halton, who perhaps believed at the time that we had departed unusually quietly and orderly, may now learn the truth. When we had been allocated rooms and issued bedding, we settled down to our new adult life.

Pat Cropley, 'Jock' Clarke and myself were separated from the other electricians, and allocated to a room for those who were employed in the synthetic trainer section. The camp had its own request radio program, and over the loudspeakers, as we made our beds down in our new home under the wary eye of our new room mates, came the announcement of "Welcome to the Fifty-Fifth entry". The particular piece of contemporary music chosen for us was, *You'd Better Watch Yourself, Bub*.

IT'S GREAT TO BE YOUNG

The sexual life of the camel is harder than anyone thinks
At the height of the mating season...

I cannot possibly even begin to describe to you the feeling of immense and total relief that came over me on leaving Halton. I had sworn to myself as I marched down Maitland Hill for the last time on the day that we left that I would never ever voluntarily go near the wretched place again as long as I lived, but that feeling of early relief was only the half of what awaited me at my new camp. Number 32, Maintenance Unit was only a half of the total R.A.F. station at St Athan, and was a huge factory complex devoted at that time to the major repair of a wide range of His Majesty's fine equipment, from aircraft engines, complete Meteor jet fighters, a vast range of electrical and instrument components, synthetic trainers of all types, some previously completely unknown, not just Link Trainers, and even gun turrets as a separate entity. It had its own huge carpentry section, an electro-plating department which was one of the very few in the whole R.A.F., and it employed not only us improvership year ex-brats, but a full complement of regular serving airmen, youngsters completing their National Service, and a whole host of civilian workers. By its very scale, this was all immensely interesting of course, but what was of far greater significance to me personally, I soon discovered, was that at St Athan there was only one real parade a week, I could wear civilian clothes off duty, there was no requirement to book out and in if you wished to leave camp for the evening, I could even stay out all night if I wished, providing that I was on work parade the following morning sober enough to stand up. I could smoke, drink, go out with girls, all without the necessary protection of any kind of chit, and, what is more, I had suddenly inherited a massive increase in buying power, now paid weekly with which to fuel it all. It was like a wild bird being released from a cage or suddenly being born again, and, as you might guess, it went absolutely and completely to my young head.

The Synthetic Trainer Section in which I was to be employed had two officers who were in nominal command at that time, but in every real sense it was being emphatically and despotically run by a Warrant Officer Charlish. He held the very highest of ranks in my Air Force, and he had long since undergone the transformation which turns a mere Flight Sergeant into something approaching a deity. The whole wide R.A.F. was at least partially his fief, having a network of all sorts of ultra-reliable personal contacts all over the place, and he was one of the few people to be able to call even God by his first name. He was simply no longer subject to the exigencies of the service that the rest of us were, officers included, and if something didn't please him then he changed it, no matter from where it had derived. When he was posted overseas, for example, the grapevine informed us of the event at once, but we didn't even have time to get out the coloured bunting and have a whip round for champagne. Just one phone call later it was suddenly all off. It would truly amaze you what he could achieve with just one such telephone call, another example of which I will recount later.

The R.A.F.'s extremely laudable intention was that we would now round out our long theoretical education with some sound practical experience for the period of a year. It was only in our individual perception of experience of precisely what, that they and I differed in the slightest. I wilfully decided to apply myself most diligently to fulfilling the very highest of the expectations of the psychiatric analyst at my intake examinations three years earlier. Being by now a well-organised sort of chap, I decided that I should start with the wine first, then progress later to the women, and finally move on to the song, but such proved the demands of the first two that I sadly found myself with insufficient time still available to develop even a modest baritone voice by the time that I was posted to Linton On Ouse at the end of our improvership year. So obvious and determined was my devotion to going downhill as rapidly as possible that it even began to alarm my good friend Pat Cropley, whose own personal inclinations had been to minimise on the drink and completely skip over the music, in order to better concentrate his full efforts on just the one remaining facet, and so do it the proper justice that it merited. He finally came to the conclusion that in my best interests he would just have to be unselfish and intervene, and take on the urgent and much needed role of being my big brother and mentor, or at my present pace I was most unlikely to last out the full twelve months.

So unselfish was his concern for me that I can only recall one single incident when we had even the slightest disagreement. We had decided to visit his home for a bank holiday weekend, and as usual had arrived unannounced, only to find that the young wife of his overseas-serving elder brother, and a girlfriend of hers, were already in residence, which served to create a small emergency of sleeping arrangements. Pat's mother, a very dear lady who always treated me as if I were one of her own sons, regretfully asked me if, in the circumstances, I would mind putting up with the discomfort of sleeping on the settee in the girl's bedroom. I didn't mind at all. As a matter of fact, I was positively enthusiastic about it, and in any case didn't really foresee any need to actually spend the whole night in any discomfort at all, but then Pat strangely cut in, volunteering to take my place and I thought I began to detect what might be described as an attempt on his part to keep it all in the family. Finally, after some protracted discourse, if not actual bickering, we were agreed that we should toss a coin for it, which we then did under his mother's incredulous eyes. She was then rightly obliged to rebuke Pat for the extremely vulgar expression he used when it was revealed that by the fortuitous choice of heads, I had won.

In his choice of extra-curricular activities Pat enjoyed certain advantages that I most certainly did not. He was a dark, curly-headed, slim boy at that time, and the girls simply fell all over him. He was extremely fastidious about his civilian clothes, and he always took good care to have at least one W.R.A.F. lady in his current retinue for the expressed purpose of washing, starching and ironing his shirts in the manner that he absolutely insisted upon. Since I began to slavishly follow his taste, but was not so fortunate as regards natural charm, and additionally, since we happened to take the same size in most things, he permitted me to simply add my own shirts to the pile that he handed over to the poor girl every week. Late delivery or objectors were simply not tolerated in his harem and were invariably replaced within the following forty-eight hours, apparently without too much difficulty. All around you it was possible to see many others of the Fifty-Fifth entry dressed in increasingly splendid elegance at all the local dances. Geoff Kent the engine fitter started a fashion in the kind of broad-brimmed hat that were much admired and worn by New York gangsters at that time. I certainly had one. Pat had much difficulty in finding a head piece that met his exacting standards in colour and style, and which would be in complete sartorial harmony with the long camel-hair overcoat that he had recently bought. The nearest that he got was one Saturday afternoon in Bridgend, when he at last found a hat of the right colour but unfortunately not quite the right shape. He began making suitable electrical fitter modifications at once by pounding it with his hard fist before the outraged and frightened eyes of the shop owner, while I looked on in the background as if bored with what was a most usual event. Finally, still not entirely satisfied, even Pat felt obliged to buy it, such had been the change to its original shape. When we finally emerged he was for all the world the absolute antithesis of his intended sartorial elegance, and more the living image of a bomb site second-hand car dealer, and once he had time to review the total effect in the mirror he rarely wore it again.

I personally soon came to rely utterly on what I regarded as his faultless judgement as regards my wardrobe. I remember the occasion that we were together in London once on leave and we saw a light grey suit in the window of a tailor's shop in, of all places, that mecca of impeccable good taste, Savile Row. Pat was not disconcerted in the least by this and his opinion was that it would look just right on me, and accordingly in we went. A very superior salesman in a morning coat, with his tape measure around his neck, looked down his long nose at us, enquired of our needs somewhat contemptuously, and when he learned that it was concerning a suit, he asked in faultless English, "And precisely what kind of material did you have in mind? Pinhead?" Pat had not significantly changed in character since Halton, certainly as regards humour, and, never one to miss an opportunity when he felt that he was being patronised, he was in there as quick as a flash.

"There is no need to get personal, surely," he replied seriously.

It was thus at the ripe old age of nineteen years that I became the proud owner of a Savile Row suit, and, more importantly, the recipient of a fundamental lesson in the height of impeccable good taste. Pat had a quite natural sense of good taste where I had none at all in those days, and whatever I learned, be it about clothes or girls, it was all from him. By dint of rehearsing me in a good line of chat, planning my wardrobe, giving me an inside track with the latest just-rejected of his string of girlfriends, which he advised me necessitated, "The gentle application of the velvet glove after having recently experienced the meat grinder approach, 'Grim'", he slowly improved my appreciation and in-depth knowledge of feminine company. By withholding half of my pay under physical threat, and even at times downright nagging, my good Fifty-Fifth entry comrade finally even managed to divert my still very extensive and rather damp social life into somewhat less extreme morning-after channels.

All this was just a preliminary to getting me back on track sufficiently to start studying all over again, whilst it was all still relatively fresh in both of our minds, and to get us an early appointment with another Trade Test Board. No longer burdened by a wing or a schools result, he had the good sense to realise that both his own and my chances were now immeasurably improved, and that our technical future in the R.A.F. was now there for the taking if we only had the wit to see it.

"The sooner we get on the promotion list 'Grim', the sooner we will get promoted," he used to advise me. Even when I was packing in preparation for going out on another servicing party job, he would come over and silently force the manuals into my hands to take with me, and, due to his efforts, I spent many evenings in distant chilly billets alone in study, when I could have been out with the others as was my real inclination, enjoying myself in some pub or at a dance. I managed to do rather well in the preliminary written examination, which gave me a bit of an edge for when we travelled together to take the oral part of the examinations. On our train trip to R.A.F. Chigwell for these tests at Easter, we found to our joy that we were being accompanied by 'Horace' and 'Dagwood', both intent on the same objective. My old nemesis of basic fitting proved to be less troublesome than I had feared, because the test job was reduced to allow for the coming seasonal holiday weekend, and I was paired with 'Dagwood' to produce a half each of what would normally have been required from each of us. Again we took our own ex-brat liberal interpretation of the meaning of a half each, and whilst I did all the rough work, for which I was most eminently suited, he completed the accurate fitting part, and so, with a little mutual assistance yet again, the four of us stood side by side before the board officer to hear that we had all been successful, and so were now in the very same position as if we had graduated in the highest rank of L.A.C. from Halton.

As a surprise bonus, after six months of service, on the day that the results finally became official back at St Athan, Pat and I heard the completely unexpected announcement of our promotion to the lofty rank of Corporal, read to us by our new section commanding officer, none other than our old friend 'Cec'. Knowing us both as well as he did by that time, I can only reflect just how much it must have deeply pained him. As we stood side by side before him, he had both our service records laid on his desk, which we had all been assured would be destroyed when we left apprenticed service to give us the benefit of a fresh start in the real Air Force. Absolutely no one can have needed that more than we both did. In the event the truth was something rather less than promised, and all that had actually been done was to black over the entries which, although preventing one from reading about the precise details of our actual past crimes, certainly showed that in both of our cases whatever it was that we had done had certainly been done somewhat unreservedly. We were both just nineteen years old. The dissidents were the ones who were on the move at long last it seemed. We were among the first of our entry to achieve promotion to N.C.O. rank, including 'Ig', and were soon followed within a period of some weeks by our good friend 'Horace'.

The privileges of the rank, such as dining separately with the most agreeable feminine company of the W.R.A.F.s, and in the privacy of the corporals' club in the evenings rather than the airmen's canteen were now ours to enjoy, and of course yet another significant raise in pay threatened to require Pat to apply the brakes to my imbibing inclinations yet again. There was no longer any need for that, however. I had come to my senses at last, and decided for myself that at this, the halfway point in our improvershment year, it was perhaps an auspicious time to turn over a new leaf, at least in my off-duty carousing, and to behave in a more responsible manner befitting my new rank and duties.

Accordingly I abandoned the licensed for the licentious, and moved into the women phase full-time, like my good friend. For openers, the young lady serving in the corporals' club was already and literally firmly within my grasp. It was she who managed, quite unknowingly, to impart to me that magical quality of respect and envy from one's peers, to which all men truly aspire. As a consequence of her late evening working hour conditions, we could not enjoy our clandestine meetings until a rather late hour, and as a direct result I was showering, dressing and carefully grooming myself in the manner of all young men, in preparation for going out to meet her, when everyone else was coming in and preparing for their night's sleep. Clad in their pyjamas and brushing their teeth, they gazed at me in open wonder in the washing rooms. It did absolute wonders for my reputation, and more importantly my self-confidence, I can tell you. At that late hour too there is little to do except cuddle together and seek private protection from the Welsh weather. There was an apt saying at the time that if you could see the English coast across the Bristol Channel then it was going to rain, and if you couldn't it was already raining. We absolutely never got wet. She kept me well supplied with toiletry products, chocolate and cigarettes too, which I shared with my faithful friend Pat, as he had generously shared his laundry service with me. There were several other significant acquaintanceships in the surrounding towns and villages, all coming along very nicely for me thank you, at varying stages of incubation, since I had, after all, to occupy my time somehow in the early evenings and at weekends, and I still wisely continued to be most diligent in playing the compassionate and sympathetic listener to Pat's rejects, who singled me out in dance halls and drew me into darkened secluded corners for private conversation asking my advice, each telling her same sad tale of rejection and asking for my assistance. I then played the part of the upright honest good fellow to perfection, just as I had learned to do in the Padre's confirmation classes a short while earlier, trying his absolute very best, but unfortunately just failing to keep his wayward fickle companion out of trouble. Pat really was extremely wasteful of talent which I would never have even dreamed of approaching in other circumstances, and any one of them who lasted a full two weeks as part of his entourage had achieved a most significant endurance record, so I was a rather busy young man. Under Pat's expert tutelage, I had most certainly improved, at least as far as the evenings were concerned.

THE WEIGHTY RESPONSIBILITIES
OF HIGH COMMAND

So like Christopher Columbus I decided to explore
And I took up my position by...

On duty, sadly things did not proceed quite so smoothly, and I must relate at least some of the incidents which beset me on my sudden heady rise to high command. The transition from being very firmly on one side of the fence to being clearly on the other did not entirely sit too comfortably with me at first, I must admit. At the time I felt that it was probably the most devilish piece of low animal cunning that the R.A.F. had played on me thus far. Their all too obvious expectations were that now I would begin to carry out at least passable imitations of any of several excellent role models of recent acquaintance that they had thoughtfully provided for such an admittedly remote possibility. Ever wayward, with opinions of my own, my inclinations were, however, to at least try and practise a daily man management philosophy more suited to the modern age majority, and save the hard stuff for those of the minority so retarded as to require it. My early experiences of applying this principle in my new role of being required to see the bigger picture seemed to be that, however I chose to play it, it was still myself who was invariably left holding the malodorous end of the stick.

Since our arrival in the synthetic trainer section, I had sometimes found myself attached to small teams of two or three, who were sent to other R.A.F. stations throughout the U.K. to service their on-unit link trainers, under the control of a sergeant or corporal tradesman. That is in fact how I had finally learned to master at least some of the necessary hand skills which I had patently failed to do at Halton, with a spot of late on-the-job-training, just as the Air Force had intended. Pat, however, seemed to be firmly anchored to a workbench in the section and was rarely sent anywhere. Perhaps our good friend the warrant officer had managed a more than cursory glance at those service record sheets of ours, and wisely concluded that a potential problem divided was preferable to one combined. Now, as a corporal myself, the working picture changed, and I was increasingly frequently required to take teams of my own out on the road, and to supervise and be responsible for them. In fact from this time on I spent very little of my improvership year at St Athan at all, and consequently was equally if not more familiar with at least a dozen or so other R.A.F. camps in the U.K., and the nature of the real Air Force still awaiting the arrival of the rest of our entry. I unhappily saw much less of Pat as a direct consequence, but, on my return from some trip or other, we would invariably be out together, he seeking new conquests, and I at least inheriting his old ones. Out on the road I applied myself to the best of my abilities to whatever happened to be available in the area.

The city of Nottingham enjoyed a reputation at that time of having a large surplus of willing young ladies mostly employed in the Players cigarette factory, and was therefore rather much sought after as a place to be sent to as a member of a servicing team. As a sad consequence, jobs in that area tended to be rather protracted affairs which tended to consistently fail to meet schedules. Warrant Officer Charlish was a very astute man indeed, however, and was probably all-too-aware of the real thinness of my charms, because he often gave me jobs on camps in that area, and I can safely say that the results were uniformly appalling. I never so much as got a single date on any of my visits. The local females invariably seemed to view me as the sheriff on a tax-collecting visit rather than any modern-day Robin Hood come to rescue them. On the other hand tiny Stamford in Lincolnshire always seemed to be a lucky place for me for some reason. Perhaps such charms as I possessed at the time were in fact more rural than urban in character.

The most demanding situations of all were the occasions when one or other of the ex-apprentices employed in the section was attached to my team, such as Dickie Denham, ex-Fifty-Second entry, and the elder brother of the Fifty-Sixth chap whom Pat had head-butted in the Halton meal queue a short time earlier. Dickie had of course departed St Athan with his entry earlier than my arrival, but, after a serious aircraft crash during navigator training in which he was badly injured, he was back as an electrician and on the point of being invalided out from the service altogether. Dickie in fact attended my wedding, so I must have made at least a half decent job of it but, although all the ex-brats now plainly regarded each other as something special regardless of entry number, I had still not quite got used to the idea. With ex-brats, I always somehow felt that I should endeavour to put up a thoroughly reliable, but at the same time relaxed, performance for both our sakes.

When I had to take 'Jock' Clarke from my own entry with me on a job to Usworth near Newcastle, it was the first time that I had been in charge of and responsible for an ex-Fifty-Fifth apprentice mate, and that most certainly made it even harder. The unit in question was not in fact an R.A.F. station at all, but was in the process of being handed over to be run by Airwork Ltd. with just a few airmen 'boggies' doing guard duties and the like in the meantime. There was only one airman cook, who lived out at his home in nearby Washington, and the food was simply out of this world, almost back to the pre-signing-on standards at Halton. Directly across the road from the camp gates, no more than fifty yards or so, was a nice country pub, and you could even pop in for a quick half in the lunch hour if you had the mind and I soon did for a very good reason. Serving behind the bar was this gorgeous creature, so absolutely breathtaking that I simply couldn't understand why none of the airmen had apparently made any move after her at all, and it certainly aroused my suspicions. Foolish I might have been, but stupid I was most certainly not. I began by making a few discreet enquiries about husbands, absent or otherwise, wrestler or weight-lifting miner boyfriends or the like, but absolutely nothing at all came to light that seemed to prevent my stepping in, so, wisely concluding that faint heart never laid fair crumpet, as my good friend back at St Athan had advised me, I did just that.

Not being so timorous and patently unsure of myself as in earlier years, and with a few practical lessons under my belt from watching master seducer Pat in action, I made earnest efforts to broaden my acquaintanceship with this young lady, and my humble efforts were not entirely without success for once. Whilst I was bidding her my very best fond goodnight one evening after closing time, in the bushes outside the pub, the officer who was in charge of the R.A.F. personnel passed by on the other side of the hedge on his way into camp and looked over, and, although he did not seem to be overly enamoured of my obvious devotion to my duty, he did not actually say anything at the time.

He did, however, say something the next day on pay parade. He left myself and 'Jock' until last, and when it was my turn to march in to his office to be paid, he bade me close the door behind me, and then told me in no uncertain terms that he had no intention whatsoever of paying a man who had behaved so despicably in front of his very eyes with his fiancée, nor could even the lowly aide of such a vile creature – from whom someone with such an obviously derelict sense of decency would doubtless borrow money given the chance – expect any usual monetary reward either as a consequence. There didn't seem to be any suitable answer to all that, and I was left to explain as best as I could to poor 'Jock', still waiting patiently outside, that he would not in fact be receiving any bounty from the Air Force that particular day, nor, it seemed, for as long as we remained at that place, for reasons sadly now beyond my control. What was far more serious than the unlikely possibility of any continuing relationship with the young lady was that the job as yet was only half finished and without money, both poor innocent 'Jock' and I would be obliged to go thirsty until it was completed. I must have been a shocking disappointment to my old comrade.

It was not always easy to be firm and dynamic with your non ex-brat former team mates either I found, and control could suddenly slip from your grasp when you least expected it. One of the first of such trips was with a good former drinking friend called Paddy McGrath, as my assistant. Paddy's outstanding virtue was that he was a very hard worker indeed, but he did like to have his Saturday night out in the local pub wherever he was and on these weekly occasions to imbibe a little more than what would be considered by most men to be his fair share. We were doing a servicing job at R.A.F. North Luffenham when the first Saturday evening came round. We went to nearby Stamford, the scene of several of my triumphs, past and future, had a few drinks together, and, then under my persuasion, attended the local dance. I was yet again fortunate and found myself an ex-Butlins beauty queen whilst Paddy went in search of further refreshment. By the time of the last waltz, she and I were rather close and seemingly of a common mind. I figured that there was perhaps just enough time to catch the last bus back to camp, and if not, well, then I might have another attempt the following morning. At that point up staggered Paddy after having plainly depleted the bar stock rather significantly, and I was obliged to make a thoroughly unpleasant choice. Certainly no one else was going to assist him to the bus in that condition. As a consequence we did not get along quite so well together in work the following week I must say, and there were distinctly long chilly silent periods between us which served to speed up progress on the job no end. In fact by the next weekend, we were returning to St Athan after completely wrapping it all up. I simply could not bear the memory of losing out at the last moment like that, nor the near certainty of something similar happening again if we stayed, and Paddy, friendly simple soul that he was, always stuck to me like the proverbial to a blanket. Our route took us back via London, and it was Saturday night again.

"Now fair is fair," as Paddy put it. "Oive done me job. Yers have had no trouble wid me, and now oi needs me recreation." I saw his point all too well. He was speaking no less than the absolute truth, and I allowed myself to be persuaded to accompany him to Mooneys, to sample the Guinness.

In the early hours of the morning I had managed to get him as far as Paddington railway station with some difficulty, but there somehow he contrived to give me the slip, and after much fruitless searching I was eventually obliged to catch the train to South Wales alone, taking both his railway warrant and his kit with me. Back at St Athan the following day, I had more time to reconsider the wisdom of my dark and creamy-headed decision, and I anxiously awaited his return all that Sunday, but without success. The following morning was a working day, and I was obliged to add in my report to the section warrant officer that, whilst the job had gone rather well, I had somehow inadvertently mislaid McGrath, and hoped that, until his obviously imminent return, he could see fit to take a lenient view of this small matter. Warrant Officer Charlish was of the old pre-war Air Force school, cast very much in the mode of the Halton wing personnel of so recent acquaintance, had never taken a lenient view of anything in his entire life and was not about to start now either. I received the rollicking of my life, and we both settled back to anxiously await the return of McGrath. He finally made it, late on Tuesday afternoon, after apparently waking up on a porter's barrow early on Sunday morning, to find the new day's passengers on the move all around him. In the absence of myself, his hat and his ticket, he had then had to go begging around the station, until he had managed to raise sufficient funds to purchase a single railway ticket to

St Athan. The period of goodwill towards servicemen normally extended by civilians in general and travellers in particular during times of emergency had apparently run out, judging by the length of time that it took him to raise the necessary funds. I tried to avoid having Paddy, good worker or not, on my team after that.

On one of my increasingly rare visits back at St Athan, I was unfortunate enough to be named as the duty corporal one evening, and it was a night when I had chanced to have what might be described as a hot date. Matters between the young lady in question and myself had lately been proceeding at quite a pace in spite of, or perhaps due to, my occasional absences on servicing parties, and it seemed that ultimate reward night might just finally be due. Such events, I might add, were happily becoming less infrequent for me under the excellent tutelage of my good friend Pat who, understanding the urgency of my position, volunteered to stand in for me on the duty, and off I went dressed up in my finery, leaving him to attend to the miserable duty.

With a joint viewpoint born of still recent hard experience, we were not so foolish as to request permission to make this arrangement of course, since that might well have resulted in an acutely embarrassing refusal. The simple arrangement was that Pat was to become 'Grim' for the night and, after all, nothing could be more simple than that. When I arrived back on camp in the early hours of the morning with the remains of a sloppy smile still on my face, Pat was in bed with his light on, unusually still awake, and I casually asked him if there had been any problems, expecting of course a negative response. However he had a very sad story for me. It seemed that he had decided to make the best of it all, and had been enjoying himself by filling in the hours chatting up the new girl serving behind the bar in the corporals' club, just for practice you understand, instead of prowling the main N.A.A.F.I. building in search of intending hooligans, or others of that ilk as he should have been. It seemed that a group of the evil ones, benefiting from the absence of any supervision, had started a small private war in the upstairs billiards room, which had finally spilled over into the main canteen downstairs and in this action they had torn down the curtains in the main room, probably demonstrating their dislike or disapproval of something or other. My good friend had heard about this turn of events rather late, arrived on the actual scene even later, and found only the damage with no one in sight to blame. He was most apologetic but there it was. Realising as well as he did from our common long experience that a very good ingenious lie was always far more acceptable, and even preferable to the authorities than the bare unvarnished truth, the following morning I wrote a long report of discovering the curtains on the floor during one of my ever-vigilant patrols, but that I had found no one with sufficient courage or moral fortitude to be willing to identify the culprit even under my hard interrogation, so I had delivered my own punishment by evicting everyone, and, what is more, closing and locking up the whole premises for the rest of the evening. The Station Warrant Officer sent for me and congratulated me on my actions the next day, and that was how honourable victory was snatched from the very jaws of defeat on that particular occasion.

Only once did Pat and I manage to get away from St Athan on a servicing job together, but you could hardly say that it began auspiciously. We had planned to use our coming weekend forty-eight hour pass to some advantage, starting with a Friday evening visit to a favourite Chinese restaurant of ours in Barry for chow fan, then on for main course to hear Shirley Bassey, who was already making a name for herself locally, and who was performing in a Cardiff working men's club, and whoever we could charm there for dessert. It all looked rather promising. Came Wednesday afternoon and Warrant Officer Charlish sent for us both, an unusual event in itself. It seemed that he had other plans for our free time, and wanted us to go to R.A.F. Innsworth, a storage depot near Manchester, with two low loader transport vehicles and their drivers, collect and remove four complete Link Trainers from their crates, prepare them for moving, load them on to the trucks and bring them back to St Athan. The task would normally have taken at least three days of back-breaking work, plus a full day in each direction just transporting them. He even chanced the merest hint of a smile at the happy thought that our weekend was now most surely blown. Small joys like that tended to make his more dull days.

We had other ideas however, and left South Wales the same evening, drove north through the night to just short of Manchester and slept in the cabs of the transporters in a lay-by, to make sure that we could legitimately claim the overnight allowance. One always had a need for a few extra shillings or so. Our airmen drivers were not entirely enamoured of that kind of spartan behaviour and dedication, I can tell you. Thursday morning bright and early, Pat and I were jointly hammering on the doors of the stores building before they had yet opened, and the rest of the day we both worked stripped to our shirt-sleeves, watched by our two petulant drivers who refused to help in what they claimed to be technical matters, and a growing small army of civilian employees who of course on principle never worked at all unless forced. They had not actually seen anyone engaged in manual labour in living memory, so we were something of a curiosity to them. When we had at last finished in the late afternoon, we were the recipients of the compliment of our lives by their foreman, who detached himself from the still head-shaking main group, came over and said, "I have always said that Air Force people are lazy, but by God you two aren't."

I didn't disillusion him by explaining the true nature of the underlying reason for our high state of activity. We then ignored the protests of the drivers and obliged them to start back there and then, overnight. Pat always had a way of honey-tonguing his way into people's trust and convincing them of the merit of his plans.

"Tired! How can you be tired? You spent the whole effing day watching us, so shift your lazy... "

Ten o'clock Friday morning we arrived back at the doors of the Synthetic Trainer Section, to a thoroughly shocked Warrant Officer, who for once could find neither appropriate words nor actions. I would of course love to be able to claim this as an example of the triumph of the boundless energy of determined youth over the bureaucracy of reactionary middle-aged establishment, but it would fall short of the actual truth. My most sincere apologies to Miss Bassey if she happened to notice, but I slept through almost the whole of her performance that evening.

HOW TO HATE SIMPLE AIRMEN

I'd find touch, she'd find touch
We'd both find touch together

Even on the occasions when you had no great affinity for your charges and could therefore more easily distance yourself accordingly, matters could still unfortunately go astray, I found. I was given a job to do at Wolverhampton municipal airport which happened to still have a very small R.A.F. group attached to it. Despite my protests, I had been allocated two airmen assistants to form a team, with whom I was all too familiar. One was a skinny teetotal Londoner named Kenneth, who supported the Arsenal, need I say more. He was a troublemaker of long repute who had once started and then involved all the members of his team, including the corporal in charge, in a pub brawl which required police intervention resulting in the eventual payment of damages, whilst he sipped his lemonade, sat back and watched it all. The other was a fat odious creature named Dudley who had, in my humble opinion, been unwisely allowed to remuster to electrician from the trade of pornographer, where such talents as he possessed truly lay. After Halton, I could hardly be described as being prudish, and was more than able to enjoy a dirty joke, and I was not and I am still not in the least narrow-minded, but his whole life and being was totally immersed and deeply concerned with every single disgusting variation of sex, and what is more, he talked about it ceaselessly. I was not exactly enjoying their companionship by the time that we got off the train at out destination, I can assure you. It promised to be a thoroughly miserable visit.

We had only just arrived at the airport, and were in fact queuing for our first meal in what served as a mess hall for the few Air Force personnel still there as well as the works canteen for the civilian workforce, when I first set eyes on the waitress, Frances. She was, to put it in a phrase, simply stunning. What was even better, I was all-too-soon to discover, was that she was tired of the local yokels and simply bored with the pitiful weedy Air Force people who had already been there for far too long, and was, it seems, in urgent waiting for some young handsome prince from afar to come on the scene and rescue her. Realising that satisfying all of these considerable needs was perhaps expecting too much even of me, I sensibly concluded that I was at least young and, although a corporal is perhaps not quite royalty, St Athan was indeed a considerable distance away if one was obliged to travel by rail, and I forgot about the rest.

It was, I came to realise in the fullness of time, a meeting truly arranged in heaven. It was so perfect that even her father, who was a devoted and most knowledgeable Wolves football supporter, by the way, and who even took me with him to watch home matches, seemed to like me and foolishly trusted me with his daughter. However, I am getting ahead of myself; all this came later. While we exchanged our first soul-melting, smouldering glances across the servery for openers that day, I remember that at one point she asked me what I wanted, and that something came immediately to mind. I was only dragged back to unpleasant reality by the mindless whining of the airmen queuing behind me, including of course skinny Kenneth and good old Dudley, that I should, "Make up your mind, Corp!" I nipped all that kind of nasty socialism quickly in the bud with just one withering look. As it happened, they need not have worried, as 'Corp' had indeed made up his mind, and the prospects for the job took on a distinctly more intriguing tone immediately.

In the fullness of time, without boring you with details, I truly came to believe that this particular servicing task might never quite be finished, at least not in the time of my eleven years or so of remaining service in the Royal Air Force. Unfortunately my two charges were seemingly neither so lucky nor so enamoured with the fair city of Wolverhampton and its charms as I. Without the benefit of the big picture perception that I, from the advantageous position of my high rank possessed, and which I did dutifully try to explain to them, they eventually began to make sheep-like bleating noises every day about, "Time we were going back!" and, "What about our long weekend then, Corp?" The ungrateful wretches even began to work voluntary night overtime on the new tasks that I daily found necessary to invent for them to occupy their time, and I would return late from my evenings out with extremely agreeable company to find them still at work in the Link Trainer room with, what is more, a light in the window, waiting for my return. Those phrases about the 'loneliness of command' came to mind all too easily on these occasions, and I began to learn to hate simple airmen. The metamorphosis that the Air Force so earnestly desired of me was beginning to take place, even if by this most unusual route.

Despite my overall good spirits and general excellence of mood, my charges began to increasingly irritate even the deeply tolerant and understanding person that I am, and it became increasingly frequently necessary to tell them to shut up and stop moaning. Such is the everyday lot of high management, I concluded with a sigh, as I continued to proceed on my most merry way. Eventually, sad to say, even my good friend our warrant officer began to miss my company back at St Athan, or, even more probably, one of the whining ones did a swift Judas on me via the telephone in my absence. Anyway, after some protracted difficulty in making any contact with me at all, because I wasn't exactly managing to keep regular office hours at the time, he finally did manage to trap me on the other end of a telephone one afternoon and gave me a rather firm deadline for finishing up the job, as well as the customary rollicking of course. After all, it wouldn't have been the same without that.

It was one of the saddest days of my life when I had to take my leave of Frances. I almost ended up punching one of my charges in the mouth in public on Bristol Temple Meads railway station for an ill-chosen remark about our parting on the return journey, such was the shocking impact on my sensibilities. When I finally arrived back in the Link Trainer section and stepped through the door, my spirits at an all-time low, deep bags under my eyes and the corners of my mouth still well down, my old friend Pat saw me, came over at once from his bench, took one look at me, frowned and said with deep non-feigned concern, "Bloody Hell, Grimbo, you look terrible. You must have had a marvellous time."

Such was the degree of my obvious antipathy to any kind of lowly airmen at that time that even my friend the warrant officer thought it prudent to give me tasks requiring only myself to complete, at least for a while. Most units that one visited insisted that you go through a long and tedious looking-in procedure, to announce to every single section on the camp that you were now one of them, even if it was to be so only temporarily, and of course the same thing had to happen all over again before you departed. It would often take days before you could get all the signatures that seemingly were necessary. This was a major headache for servicing parties on short visits, and I personally tried to avoid it whenever I could, and working alone made that infinitely more possible, particularly when the camp or locality was less than inviting and I could make the trip a really short one. At such places I tended to be in and out like the proverbial through a goose before anyone even knew that I was even there.

Our Warrant Officer Charlish, before he came to know me just a shade better, had begun to look on this very brisk solo job completion rate with some favour, and seemed to interpret it at the time as a certainly belated but still highly welcome inclination on my part to be dynamic, a general liking for hard work not usual in my generation, in his opinion, and perhaps a desire to even further advance my career, and that probably had at least something to do with how I eventually came to find myself on a commissioning interview, come to think of it. In view of what had previously happened when he let us loose on the R.A.F. together, our leader was seemingly no longer prepared to let Pat and me be in each other's company for too long, and the sooner that I returned from some job or other, the sooner I tended to be out on the next one.

One Saturday morning, I found myself at R.A.F. Benson in Oxfordshire, getting a sign-off from another satisfied customer for yet another completed job, when the officer signing my work sheet told me that he was flying an aircraft back to St Athan for a major service within the hour, and asked if I would like to have a lift back with him. Until that moment, all that I had considered as my likely immediate lot was the choice of either a dull weekend alone in the country or another long train ride back to Wales. Immediately visions of a more genteel and spiritually fulfilling evening out with Pat in Cardiff, followed by a long Sunday lie-in came to mind, and I most gratefully accepted the offer. We had landed by ten-thirty, I was through the doors of the synthetic trainer section by eleven, waving my thumbs up greeting to a surprised Pat as I gaily steamed past him into the office to report my return, and I emerged somewhat chastened, issued with a new job sheet, railway warrant, and booked out on my way to R.A.F. Horsham St Faith in East Anglia by twenty past. Mr Charlish had long ago mastered the art of handling all forms of the unexpected including the return of prodigal sons ahead of time, you see. Of course I didn't actually depart that same morning or the following one either, but I was obliged to resort to the indignity of hiding for a while in my own clothes locker, and sneaking down to the railway station the following Monday morning after everyone else had gone to work, but my warrant officer was certainly not taking any prisoners in his dealings with me.

There was a darker side to his trust and confidence, however, I must add, and he also took to entrusting me with small extra duties of responsibility and confidentiality, the nature of which did not always greatly please me, on the few short stopovers that I did manage to have at St Athan. He sent for me one morning and confided that a certain Senior Aircraftsman Hacker who was employed in our section had requested permission to be allowed to proceed to St Athan East Camp that sports afternoon to witness an inter-service swimming gala going on there, rather than taking part in any games himself. He was not, it seemed, entirely confident of S.A.C. Hacker's true intentions, which just might be to catch the next available passing bus and go and visit his nearby home town instead. My instructions were to follow at a discreet distance and closely observe the potentially errant Hacker, and to nip in and prevent any such travel arrangements if that awful scenario looked like coming to fruition.

At the morning tea break the plot further deepened when Pat came along seeking me out, took me on one side for a quiet word, and imparted the information that he had apparently now been admitted to the inner spying circle too, and he had been detailed to follow me following Hacker as his first assignment. We privately agreed that perhaps we could both satisfy our murky responsibilities to our master by at least discharging our unpleasant duties in mutually pleasing company. Neither of us was an expert in this particular field, I hasten to add, being far more familiar with the role of the fox rather than that of the hound, but we managed to maintain what we both felt was a suitable spying distance, and jointly stalked poor Hacker down the back road to East Camp that afternoon, but we must have been something less than skilful because finally be stopped and waited until both Pat and I had caught him up, and asked us just what the hell we thought we were playing at. Once the three of us had taken the elementary precaution of making quite sure that there wasn't anyone following Pat, we all agreed that the only remaining matter to be settled was if we should all watch the gala or all go to town to enjoy an early afternoon off. Since Hacker was obviously a great home lover and both Pat and I had current girlfriends living in the town, we jointly decided that even remotely possible sex was infinitely more attractive than swimming, and made our decision accordingly.

In spite of my very earnest best efforts such as those, however, it was not too long before the high opinion of my humble talents that I had quite inadvertently aroused within my warrant officer was severely downgraded to its original level and even lower, due to my apparent unfortunate ongoing tendency to be accident prone. He did, however do me, or was it himself, a favour on at least one later occasion that I remember.

I had been called in and informed by 'Cec' one day that I had been selected to attend the next course for training as an inspector in the Aeronautical Inspection Service, the group responsible for inspecting the work of others rather than carrying it out themselves at maintenance units. He seemed to think of it as a privilege of some kind, but I didn't like the idea at all. It probably meant my being employed exclusively at maintenance units for the rest of my future service in all probability, and my heart was already firmly out there on the flying squadrons. I worried about it for a few days, before wisely concluding that if I didn't yet know how to get round this kind of problem myself then I most certainly knew who did. I wisely paid my warrant officer boss a visit and explained my reluctance, for once with honesty, since I couldn't think of anything better. It was most certainly all news to him and the concept obviously appealed to him as much as it did to me, realising that it would probably be him who would inherit me on a permanent basis in that eventuality. It took him just one phone call there and then on the spot to get it effectively stopped, to the vast relief of both of us.

Perhaps I did just once knowingly contribute ever so slightly, to the ongoing deterioration of the level of esteem in which Mr. Charlish held me. As far as any of us mere mortals ever learned he had only one small weakness, and that was an avid interest in the sport of tug of war. He was the organiser and trainer of a station team which had only one serious competitor in the whole R.A.F., and was the frequent winner at all local sports events at the very highest level. From mysterious sources he could, and did, arrange for such unheard of things as pre-match sherry and eggs for the team, and special high protein diets. Many of them came from the ranks of the synthetic trainer section, and, in his enthusiasm, he was sometimes given over to totally abandoning normal work for the whole section, getting us all outside on the grass and obliging the rest of us to assist in the team's training by forming the disorganised opposition on the other end of the rope. He was simply ecstatic when they dragged us around on our uneducated and unenthusiastic bottoms. I, amongst several others, tired of this, and consequently there came a day when we had organised ourselves somewhat better beforehand. The heavy beef amongst us took up their places at the rear anchor position with instructions to just hold, and the more whippet-like and muscular settled down at the front, with instructions to wait until a deadlock position had been established, and then on a quick word of command from one of us, go to a snatch and jerk mode of action. On that occasion it was we who dragged them around. Our leader stormed back into his office red-faced with anger, but when he had cooled down a trifle it didn't take him long to conclude what was the most likely source of any kind of organised anti-establishment action such as he had just witnessed. All the ex-brats, regardless of rank, suddenly found themselves elevated to prominently high positions on their individual unpleasant duties lists. Even never-go-anywhere Pat was being dispatched for a while to the joys of R.A.F. Kinloss, far away in Scotland, and I found myself bound for R.A.F. Aldergrove in Northern Ireland, cabinless on a crowded overnight ferry and in extremely rough weather. Mr Charlish could apparently also arrange that sort of thing.

I was dispatched on one such single man trip as I have earlier described to R.A.F. Valley in Anglesey, North Wales, and my mind was made up before I even got through the camp gates. North and South Wales are not really very far apart geographically, until you try to go by train that is, usually via Crewe which can only add to the overall charm. On this occasion it took me from early in the morning to quite late at night to get there. From the moment I stepped off the train, I encountered major mega-type problems in finding anyone even willing to speak English, let alone supply me with basic directions, and it was with great difficulty that I finally made it to the right bus to get to the camp. Accordingly I decided to make it a rather brief first visit, in view of the strength of welcome, and got the keys to the Link Trainer building from the guardroom as I booked in, and was informed that it was located on the far side of the airfield. I was somewhat peed off, to put it mildly, as I set off into the pitch black, with the serious intention of working overnight and getting the hell out of there the next morning, and perhaps that may have blunted my judgement a trifle.

Finally I did find the building, after almost straying on to the main runway with night flying in progress on a couple of occasions, and I switched on the building lights and started to check over the trainer and put it through its paces in a distinctly black mood, with a growing thumping headache to match. At one point in the early hours of the morning, I had to go outside to check the turbines which had been located in outside sheds to diminish the noise. It was a perfectly foul night, or rather early morning by then, raining cats and dogs and with a gale-like wind blowing from off the Irish Sea. Suddenly the door went slam behind me, with the keys still inside my jacket, which, by sod's law, was inside the building naturally, and suddenly there I was on the far side of the airfield, all alone in my shirt-sleeves and trousers enjoying the balmy climate. I was pretty well soaked by the time I had found a sufficiently large stone, broken a window and got back in, plus I now had to remain the next day to report and explain the broken window to the officer in charge, who had to sign my work sheet. He was not impressed with either me or my obvious devotion to duty; in fact, he plainly thought me quite mad, and I ended up having to pay for that window before I was permitted to leave, and, to make it worse, my conduct, can you imagine, was reported back to my good old friend Warrant Officer Charlish, he of the lenient viewpoint. So on this particular trip all I had to show for my efforts was yet another jolly good rollicking, a bill for a window, and a rapidly developing dose of flu from the soaking that I had received, so you can perhaps understand my cynical amusement when I have sometimes seen advertisements bidding one to visit and enjoy the delights of 'Sunny Anglesey'.

THE PARTY'S OVER

And when they went back home again
There were four and twenty less

Although I continued to bounce like an errant pinball from obstacle to obstacle, and trouble to trouble, seemingly never failing to make at least some of the lights flash and the bell ring, and with the visible count of small demeanours rising steadily, perhaps it is sufficient to say that, through many incidents like those that I have described, I finally grew up just a little during that year, both as a regular serviceman and a man, and in the end I did not have to compromise myself too far in doing so. The R.A.F.'s very laudable intention of trying to give us all a tougher practical edge in that improvership year was not entirely successful, in the case of the electricians at least. Our trade simply covered far too wide a range of possible demands to make that even remotely possible, and was soon after divided into electricians (air) and electricians (ground) in a failed attempt to remedy this, although we of the old school perversely remained electricians (both) even when there was no longer any such trade, but that particular piece of inconsistency never seemed to bother the Air Force too much.

'Horace', for example, working on aircraft in the Special Installations Section, was by now very familiar with the problems involved in working on several types of totally unfamiliar aircraft, but had little concept of the skills required to service link trainers, and about bombing or gunnery teachers he knew absolutely nothing at all, whilst I did not yet even know how to get a cockpit cover open as a preliminary to tracing some fault, or even change a set of batteries, but I did know a great deal about many types of synthetic trainers. In the end it is experience that is the best teacher of course, but it can be painful, particularly when you are suddenly cast in a supervisory role, and yet your own self-respect demands that you must be at least as good and probably better at any task than those for whom you are responsible.

Then one day the year was suddenly over and the moment for which we had all been prepared was at hand. There was a farewell dance at St Athan East Camp, and a final opportunity to say your goodbyes, even to your comrades. The next day I had the privilege of taking Bill Mercer, the horticulturist and pruning knife-wielding 'plumber', and two others of my entry north to R.A.F. Linton On Ouse to join the flying squadrons of fighter command, whilst Pat took others south to the experimental flying station at Boscombe Down. 'Horace', always aiming for the very top, had somehow in his improvership year contrived to arouse the personal wrath of no less a personage than Air Chief Marshal Sir Basil Embry, the Chief of Fighter Command, who had sworn that 'Horace' would never be allowed to as much as touch any of his Fighter Command aircraft. Well, with typical Air Force consistency, 'Horace' was now on his way with Johnny Hopper to Middleton St. George, to work on Meteors, the front line fighter aircraft of the time of course, what else. The real Air Force was awaiting us all at long last. The rest of the electricians of my entry had already largely moved into their own different worlds, and I had seen little of them since Halton, nor, as it transpired, was I to do so in the future, although I met many of the electricians of other entries over the years. Occasionally though one did hear about one or other of us from time to time, as the Air Force seemed to grow smaller through the years.

As a preliminary, my three comrade charges, with quite characteristic disrespect, threatened to lock me in the toilets if I didn't mind my manners, as they had done to their last corporal minders on the move to St Athan just a short year before. That sort of thing helped to keep your feet on the ground if you chanced to begin to get in the slightest way big-headed. Privately anxious not to disappoint them, I used the usual mixture of low animal cunning, well-told lies and a few other skills acquired over many journeys to many camps during the previous year to arrange for transport to come and pick us up – rather than suffer the usual advice to try to squeeze yourself onto a bus with full kit – and a waiting hot meal rather than a churlish, find-your-own-in-the-canteen when we arrived at York. Their cynical and appreciative smiles, whilst perched on the edge of his desk, as I went through my act to the camp's orderly officer over the stationmaster's non-paying telephone, said just about everything that need be said about our common relationship. They seemed to particularly enjoy the bit about how we were 'safeguarding and transporting an officer's private kit'. Poor 'Cec,' so recently left behind at St Athan, still had his uses, even in memory it seemed. We then neatly stacked the only kit we actually were carrying, our own, and repaired to the pub across the road for a quick pint or two together, thoughtfully leaving a message with the ticket collector on the gate as to where the 'boggie' duty driver might find us when he finally arrived. The education of the previous four years had not been entirely wasted on any of us you see, and we now knew how to make the service work for us on occasion, rather than always the other way around.

A LAST GLANCE OVER YOUR SHOULDER

Hey jig a jig, hey jig a jig, follow the band
Follow the band with your...

I had come into the service barely four years earlier with a feeling of deep anger, and at the same time a need to belong again. Now any anger was largely abated, but the little that remained had most certainly located a far more tangible entity as a focal point. I had indeed found a new family to which I very much belonged, but now it was breaking up it seemed. At the time when I had at last come to be where I wanted to be and to finally start being my own man, I realised that I had in fact come full circle, and my uppermost feeling at that moment and for quite some time afterwards was not, as one might have expected, joyous eventual fulfilment, but a sense of isolation, loneliness and loss. It was a quite short journey from my camp near York to my home town of Grimsby, and I could have gone there almost every weekend had I chosen to do so, but, after just a couple of touching base visits, I rarely did. If Halton had taught me anything at all, it was that I most certainly did not belong there any more. I was beginning to acquire something of the regular serviceman's art of making home wherever you happened to be. Without the others around to stimulate and encourage me to merge into the local social scene, however, I seemed to lack much real interest in anything at all for a while. I was acutely aware that, when the day hadn't gone particularly well at work, and in a completely new and unfamiliar technical environment there were many such days at first, there was no longer a Pat or a 'Horace' to tell it all to, just myself.

Looking back on it all, over the years, my abiding memory of Halton is of marching back to number three wing en masse on an autumn evening with the sun going down, and the lights already visible in the windows of our barrack room home, silhouetted against the Chilterns, after yet another hard day at workshops, a little tired and downcast in spirit, with so many long hard days still ahead of us. As we would crest the hill, the pipe band out in front, sensing that general mood, would sometimes change tune to 'The Bear'. At once the change could be perceived in every one of us. All around you the shoulders would lift and the heads would go up. At that so special part of ours, the combined shout would go out as one, calling out our mass defiance of the system to the whole world, and our sheer pride in ourselves. We truly were something very special then.

The last time I heard that particular tune being played at least in part for me was at that farewell dance the night before we were all leaving St Athan on our way into the big real Air Force. It was well populated by the last entry of apprentices to undergo an improvership year together and their young ladies. We had all improved considerably in that direction too in our short year, judging by the quality of the young females on display. After the ballroom had darkened and the traditional last waltz had ended, even as last furtive farewell kisses were being exchanged, from a darkened corner a solitary piper started up with our very own tune. From every corner of that dark room I heard every Fifty-Fifth voice present join in for the very last time that we would be together as one. We had played ourselves in to it just a short year before, and we were going out to it too.

A BRATS MAFIA MADE MAN

The next port we called at was Aden
The girls wouldn't

Time eventually took care of my rather sombre mood of that time of parting as it rolled on, taking me along with it and little by little adding some extra dimensions to my life. I was very fortunate, I now realise, and in the fullness of time I did get to live my dream of being out there on the flying squadrons, accepted and respected as an equal, but, like all dreams, sadly it was only for a little while. I had already been married for over a year when the imminent arrival of my baby son delayed the posting to the Far East, for which I had apparently volunteered by my failure to step backwards at the appropriate time, and it was after I returned in 1955, and incidentally at last had some legitimate reason for being in the sergeants' mess other than polishing the floor, that I got down to some serious study, and managed to get myself a Higher National in electronics, to go with a previously obtained Ordinary National in electrical engineering. If I had to point to a time when Halton had finally made me into some sort of end product then it would have to be then when I arrived back in the U.K. complete with suntan and a permanently changed viewpoint on life. Then I felt, and with some justification, that there was absolutely nothing at all that I couldn't handle. There was the occasional reminder that the Air Force was still not entirely consistent in its dealings with all of us, but that sort of thing only amused me by then. While my last three digits of service number were 602 and Pat's were 645, and bearing in mind that not all of the 43 intervening numbers were electricians, since we had both enjoyed precisely parallel careers, even being promoted on exactly the same day, and had volunteered for overseas service together, one might have expected that we would be out of the U.K. doing our overseas tour at approximately the same time, but no. Even with my delayed departure, I had long been back before he even went.

Throughout all my long disenchanted service I was being sustained and supported at odd unexpected times by apparently being viewed as a fully paid-up member of an exclusive intimate group, recognised mainly by ourselves, with no concern at all for entry seniority now. You could never be quite sure either, when it might suddenly and most unexpectedly come to your aid.

On a trade test for promotion to Senior Technician held at Seletar in Singapore, I found myself being questioned by an officer who was most certainly already a life member of the Round, Brown and Hairy Club, more than anxious to demonstrate just how well he could read from the manual held on his knees beneath the table which separated us, rather than finding out how clever I might be. He demanded that I draw the circuit diagram of a certain complex piece of equipment from memory and, satisfied that this was nigh on impossible, he departed to get on with other important matters befitting his rank while I got on with it, leaving me under the watchful eyes of his warrant officer fellow examiner. When he had left, the warrant officer picked up the manual from the chair of the officer, still open at the page containing the diagram, placed it on the table facing me and went to the window with his back turned to me and studiously remained gazing out until I gave a discreet cough. When I departed Seletar, having successfully passed the trade test, I went to the warrant officer's office to shake hands and so express my thanks for being allowed to get over that piece of utter nonsense. Only then did he enquire with a smile what my entry was and reveal his own.

It didn't always exactly work in your favour however. Over tea in the sergeants' mess one Saturday evening, I was once foolish enough to accept the invitation of 'Ginger' Fraser, ex-Fifty-First entry and former apprentices' cricket captain, to accompany him in his car for, "Just a quiet drink in a nice country pub that I know," that evening. 'Ginger' had received the benefit of a private education before Halton and his English and general demeanour were quite impeccable. Unless you looked more closely at the eyes behind those rimless spectacles of his, that is.

Mine host of the pub was very much a hunting, shooting, fishing devotee judging by the decor. "I say, landlord," 'Ginger' began as we took the first sip of our pints. "That is quite a remarkable stuffed pheasant that you have on display there behind the bar. Might I perhaps examine it more closely?"

The landlord was only too happy to oblige him and 'Ginger' examined the glass case closely both with his spectacles and without and from all angles for some time. Then he asked, "Do you think I might have the glass cover off too?" and all the while bubbling on about what a wonderful specimen it was and the like. With the cover removed, the landlord was obliged to go down the bar to serve another customer, at which point 'Ginger' took his cigarette lighter from his pocket, lit it and held it beneath the now exposed pheasant which, full of stuffing and preservatives, went up like a rocket on Guy Fawkes Day.

"Come along Roy, my dear fellow. Finish your pint," 'Ginger' said. "The beer here is quite appalling and perhaps its time that we took our custom elsewhere."

Sometimes it was our old friend authority which became the target of the never-failing mischievous spirit inherent in most of us, when it suddenly chose to re-emerge. When sports parade was reintroduced at a training unit by a wing disciplinary warrant officer who was a most obvious throwback to Halton days, it was demanded that even senior N.C.O.s should parade, complete with their intended sports equipment. Thus it was that Sergeant 'Dave' Davey, ex-Fiftieth, stood frozen-faced next to me on the parade ground holding the bridle of his fidgety horse which on evidence was certainly not yet completely broken, whilst I firmly grasped my snooker cue in the 'at ease' position. The very same warrant officer was dining at the same table as myself in the sergeants' mess one evening when we were joined by 'Ginger' Fraser at his friendly best, who chose to sit down next to him. I might add that 'Ginger' had always enjoyed a most remarkably good appetite. They had made their selection from the mess waiter's menu and both had just been served when it began.

"I say, sir," 'Ginger' opened, "what an absolutely splendid Mess Ball it was last evening, don't you agree?"

To the grunted reply in the affirmative he then went on, venturing to go so far as to dig the warrant officer confidentially in the ribs with his elbow, as if they were the oldest and dearest of friends.

"I saw you, sir," he chuckled. "You really are a sly old dog, you know. Tell me, where on earth did you find that old scrubber who was with you?"

The angry roar that went up focused everyone's attention in the whole dining room and reduced it to a stunned witnessing silence.

"Sergeant, how dare you? That was my wife!" he bellowed. 'Ginger' was then full of the most profuse apologies of course, but the warrant officer, all red-faced and angry, got up and stomped out, totally abandoning his evening meal. 'Ginger' then smiled at me happily across the table as he continued, "Well, he won't be needing this then," and he filled his own plate from the now departed warrant officer's.

This barely concealed disrespect for authority never wavered for an instant. After a pre-dawn readiness exercise programme on a Far East fighter squadron, the aircraft were finally armed, fuelled and ready at first light, and the sweating oily ground crews, including an aged corporal ex-brat engine fitter, were sat on the ground outside the section offices awaiting the more civilised hour arrival of the aircrew from the officers' mess. When they did finally grace us with their presence and dismounted from their truck it was seen that they had been armed with revolvers for this particular exercise, but there had obviously been insufficient holsters to equip them all so several, including a very youthful cherub-faced pilot officer, had tucked the weapon in the waistband of their shorts as an acceptable alternative. The fitter looked at them, now closely approaching, in obvious complete disdain.

"Look!" he said in a loud voice for all to hear and pointing a greasy finger at the young officer "An effing juvenile delinquent."

As the end of my service drew nearer on my last station, I felt it was time that the Air Force granted me permission to occasionally drive their motor vehicles, not for pleasure but as an urgent means of carrying out my pressing daily duties. My section officer, however, was not inclined to grant my request, nor on general principle to make any decisions at all, for which he might later be required to offer any explanation to higher authority. How I overcame the practical day to day problems was in his view my own business, about which he did not wish to be informed, and I was therefore obliged to make my own arrangements. The ground electrical section of which I was now in charge were urgently modifying all the many Land Rovers in the unit at the time, and the building was only large enough to contain one vehicle at a time. The vehicles were delivered and left outside our doors in groups as they were available. Rather than endlessly wait for a driver to arrive and take the completed vehicle away so that the next one could be brought in, or be reduced to pushing, I used to drive the vehicles in and out of the section so as to speed up the process. I must admit that it was usually carried out with a good deal of shall we say elan, with some hard acceleration, squealing tyres and violent braking which was a great morale booster for my faithful staff, particularly on rainy days. Unfortunately our noble leader who had refused the permission chose one day to be turning the corner on his stroll for an intended visit to my humble outpost of his vast empire as I shot out of the door in reverse and neatly pinned him against the wall, with ample inches to spare I must add. I climbed out nonchalantly and gave him my customary best barracuda-like reassuring smile. The only consequence was that when he had accepted my kind hospitable offer of a cup of tea, sat down in my office chair and his hands had finally stopped trembling, I was advised to try my very best to restrict my driving to urgent emergencies only, and to above all avoid letting the Wing Commander see me under any circumstances. It was some time before he graced my section with another visit, but only a short time later, when he was himself driving a Land Rover around the perimeter track of the airfield in a leisurely manner, I overtook him at high speed behind the wheel of a ten ton Coles Crane. I again gave him the best friendly toothy smile again, and he decided to await my successor more anxiously.

Finally, on October 1st 1960, an only marginally older but considerably wiser young prince finally made his exit stage right from the Royal Air Force without the benefit of any remission of sentence for good conduct, which was entirely as it should have been in all honesty, because there had not been too much of it worthy of note. It had been agreed before we left Halton that there would be a reunion on what other possible day than the 5th of May 1955 on the steps of St. Paul's Cathedral. I was half-way round the world in Hong Kong by that time, so I don't know how well attended it was. Some of those smiling hopeful boys from my entry who had volunteered for aircrew, including my fellow electrician Ken Smith and my overseas serving comrade 'Chas' Tighe, were already dead by that time, killed in flying accidents during or shortly after training. Many years after leaving the service and being then effectively separated from any source of further information, I often enjoyed the odd private, nostalgic evening moment, saw those young laughing faces emerge from the far corners of my memory and thought how nice it would be to see all the survivors again and to hear how life had treated them, because they were the very essence of what made the whole experience worthwhile for me. However, I could not and still cannot bring myself to look back and happily say how wonderful it all was in those times of our youth, if in doing so it might give even tacit approval to that to which I am still implacably opposed. Essentially it is the same old 'Grim', you see.

We are certainly all out of the Air Force by now, even those who chose to go on for a career to the age of fifty five or even later and somehow managed it. I believe that 'Horace' was the last one of us in uniform after some forty-one years of continuous service. Probably many had no other possibility than to go on when they left Halton, let alone by the age of thirty when their first engagement was ending. I had one thing very clear in my mind as I left Halton. I was going to leave the service the moment that my engagement was completed, and in the meantime I would have to study and plan in order to make it possible. I had learned enough already to believe that I was capable of better things than the Air Force was probably ever going to be prepared to give me.

HINDSIGHT

*This is my story, this is my song
I've been in this Air Force...*

These days a little older, and hopefully also a little wiser, I try to be realistic rather than purely vindictive, and even occasionally appreciative rather than merely abusive about the things that Halton may have done for me. The reader by this time must have gained at least a small insight into the fact that the R.A.F. and I were not always entirely of the same mind during those formative years of mine, and perhaps it is sufficient to say that it never did materially change with the passage of time for either of us. Looking back, Halton, it seems, mainly taught me the dimensions of what it was to which by my very character, I continually just had to be opposed, and I had to educate myself yet again in just how to go about it all, but I didn't always quite succeed in doing that either I am afraid. From time to time, right until the very end, the service would seemingly always just have to try to remind me of who they felt that I was and what was my place, and so rekindle that anger.

I well recall the day when I was stood before a young education pilot officer about a year before my engagement was due to end, asking about the possibilities of an early release to take up a university place then on offer to me. I had, by that time, I felt, long paid my dues by anyone's standards, and been a member of the Air Force, albeit my one, considerably longer than he had been a member of his, and I had not been entirely without success at it either. He looked me up and down in much the same manner that the civilian tailor had done at Halton all those years earlier, and replied that he did not think I was the university type. As far as he and the R.A.F. were concerned, that was effectively the end of the matter. By that time I had learned, however, how to be a little more devious in my outspoken opinions, and I did finally manage to have my say on this matter, although I had to wait for some months for the opportunity. When Christmas arrived and, according to custom, the officers' mess were invited to be the guests of the sergeants' mess for a pre-lunch aperitif, I found myself serving drinks to this young officer among others and he apparently was no longer mindful at all of our earlier conversation.

"I say, sergeant! I hope you are not one of those sort of fellows who slips a scotch into an officer's beer in the hope of getting him drunk," he laughed.

"Oh no, sir," I replied, "you can absolutely trust that I would go straight to the cyanide."

"Hah! Hah! Cyanide," he giggled thinly, noting that I was not even smiling. As he sipped his half of ale nervously, I ventured to continue confidentially. "Tell me sir, since it is the season of goodwill and all that sort of thing, if you had been bright enough to attend university, would you have gone?"

He didn't stay for his free red balloon, packet of sticky sweets or piece of Christmas cake, nor did he even bother to finish his beer before departing. He had a pressing engagement elsewhere it seemed, probably with Father Christmas in whom he obviously still deeply believed.

Halton certainly made me independent, neat and tidy for example, and taught me to take a pride in myself that goes far beyond mere appearance, to be self-reliant and resourceful, to be aware of my capabilities and limitations, and sometimes even to go beyond them when I have to. It most certainly made me aware of some of the harsher ways of the world, and confirmed the variable nature of human beings, even those close to you. It even finally taught me to keep my overly active mouth just a shade more shut on occasion, to keep my private thoughts just that, and to act on them when matters eventually turned in my direction in the fullness of time. The things that it really taught me, however, were ones that were perhaps never truly intended: it taught me belief in myself above orders and to not unquestioningly accept authority on face value, and that my instinctive value of myself is far more important and reliable than any organisational value. It taught me not only how to survive in difficult times, but also the need to resist and how to resist when you have very little with which to do it. It taught me the value of friendship, sharing when there is nothing left to share, the need for some compassion for others, and for individuals to band together when faced with injustice. In all my subsequent service, and from very serious intentions, I never treated anyone as badly as I had seen people treated there on an ongoing daily basis.

The truth is, of course, that I particularly, amongst several others, was totally unsuited to that kind of life and to those sort of values. Some, like me, were fortunately later able to discover their own routes in life, and their own ways of finally fulfilling what they essentially were. I understand now in maturity that each one of us separately brought his own particular offering, as well as personal kit sufficient for five days with us that first afternoon on Wendover Station and Maitland Square. Who knows, maybe all that I brought was my still very deep sorrow at what God, for want of a better word, had done, not to me, but to someone that I cared for above all else, and who deserved far better. In the most unlikely event that, in the words of one of our entry songs, I am ever so fortunate as to go to heaven, then I am afraid that my first act will just have to be to take issue with him that he in fact made a very serious mistake. Others less disturbed brought a very deep, sincere, and their very first, love to a service that they still didn't really know and, make no mistake, many never lost that deep affection for the R.A.F. and I truly and sincerely respect them for it.

I came to realise that the service which I believed I was joining didn't exist any more, if indeed it ever had. As the war had ended, the career men had crawled out of the woodwork and taken over where they had left off. I am ashamed to say that even some of my own entry, who should have known far better, took this easy route in later years. I also saw real embryonic leadership qualities not just ignored but deliberately suppressed and wasted by those who were all too aware that they were unable to match them, and I saw what results you get when you give free rein and authority to the moronic and the petty, who have no real values at all. It wasn't always the individual who was the loser either. I saw the service discard real technical ability, which it most desperately needed, when it found that it couldn't subordinate it in the way it liked.

As at least a partial consequence, although I must admit that I did have something of a head start, I tend not to like, trust or respect authority, at least until it has proven itself to me. It is invariably wrong for someone, and that someone can sometimes be me. Neither do I have any great liking or respect for blank obedience and imposed discipline either. The only discipline really worth anything at all in my opinion is self-discipline and, contrary to Air Force thinking, you can't always teach that by the over-generous application of the other kind. I saw entirely too much of that then and later. The truth is of course that there are just too many self-seeking managers in this world, both in the service and out, and so very few leaders. Leadership stems from a common respect and a common purpose, not orders. I learned to rather carefully watch the eyes of the one giving the orders, and to detect when it might not be in my own or the general best interests to follow them. That way I learned something about survival. If there is a purpose in life at all, then it is to try and be a decent human being while you are here. I don't especially recall that being on the curriculum at Halton.

Looking back at the total education that we received as a group, I find myself surprisingly positive about it all. For the greater part, the shall we say character-forming part of our education, which I absolutely loathed and despised at the time, would seem, in the overall quality of most of the end products with whom I am in contact, to have been remarkably successful, particularly in the context of prevalent present day social behaviour patterns, although in the case of one of us you might get a significant argument from my wife on that score. The R.A.F. always tends to take great credit to itself for that, but I believe that the truth is at least partially something else, and that they were just the singularly fortunate beneficiaries of a general strength of character which in most of our generation and earlier ones was very largely already in place when we arrived, and any credit due for that belongs elsewhere. The hard discipline and blind obedience approach applied to all of us was of a by then already bygone age, and showed at least a certain lack of grasp of the fundamental changes that had been taking place in society since the early thirties, and the nature of the new generation, of which I was a part, who grew up during the war years. Without any doubt we assuredly needed some sharp curbing and controlling of what certainly were collectively rather high spirits, but what we received was something far in excess of that.

As to our schools education, so much more could have been achieved if the R.A.F. had only got its post-war act together and the early teaching had been to the same standards as that provided by just one outstanding, dedicated individual in the case of the electricians. Even with these limitations, I must honestly say that it was still possible to go on to further higher education in later years and not to feel too disadvantaged or disgraced in the company of younger classmates who had arrived by a somewhat easier and less tortuous route via civilian education classrooms, and who had probably been able to receive their rock cake without milk too, come to think of it. You just had to be slightly more determined and work a little harder to get there, that's all.

On a more practical level, despite what the R.A.F. and other wholehearted supporters of the apprenticeship system may choose to claim, I certainly assert that the total trade education which we electricians received at Halton, although perhaps adequate to carry out our intended service by replacement roles with the equipment then in use in the R.A.F., and deep enough in background to be able to use our improvisation skills, still lacked vision about the direction in which technology was moving so rapidly, and in the final analysis, this has eventually suited the modern R.A.F. no better than it suited us.

In my civilian times immediately after service years, when I was often sat at my junior designer's desk, sucking my pencil and seeking inspiration, or trying desperately to construct a prototype that at least worked, I often painfully mused that a technical level in electronics which only went as far as the pentode valve for apprentice electricians had hardly managed to equip me for a life like this. In a less personal and broader general sense, the sort of skill levels in this area that we had been taught were in fact no longer relevant by the time that our first engagement was ending, only ten years or so after we left Halton, yet computing and electronics had probably been the major growth areas of the electrical industry in the same time period, both inside and outside the service. Unfortunately too our training was not quite comprehensive enough to enable the majority to adjust easily to earning their living on something approaching equal terms after leaving the service. Apart from defence industry companies or civil airlines, many found that the best positions available to them were jobs as technical sales representatives in civilian life, which is hardly at the sharp end of technology, yet they most certainly did not lack technical ability. The thing that always overcame this to some extent, it seems, was that old sheer pride in ourselves and that ability to improvise with very little so, perhaps, in spite of my protests, the system did manage to triumph after all, even if what worked was very much one of the learned and not taught subjects. Although subsequently some extended their service, only two of the electricians went on to eventually take technical commissions, neither got terribly far and most found that their career lay outside the service altogether. That I find truly wasteful and highly regrettable.

Despite apparently having been trained to more exacting standards for what we anticipated as being an intended eventual faster track and higher role than that of direct entry tradesmen, when we finally got out into the mainstream Air Force we found that in fact there was no higher role and virtually no distinction being made between us either. Halton was still sublimely going on attempting to teach for an earlier age in the service when any group one tradesman really was something special, and a Leading Aircraftsman, with his eighty per cent minimum trade test pass level, a precious asset to any unit because of his absolutely guaranteed technical competence. That time, however, was already long past in practice, and the service was in fact progressively moving in the completely opposite direction, repetitively sub-dividing and watering down standards in order to make qualification easier, just as the English post-war education system has done and with the same abysmal results. Many found that their trade no longer even existed or was so reduced as to be virtually worthless before many years had passed, and they then had the choice of quickly adapting to learning a new one, leaving the service altogether, or the ultimate extreme of considering a fresh start at the bottom of that other Air Force.

After reaching the rank of sergeant relatively quickly compared to some of the other fitter trades, the ex-brat electrician was virtually at the end of the line. In my time of service, I only actually saw one promotion from sergeant to flight sergeant, and one from flight sergeant up to warrant officer, both very near the end of their full careers and neither was an ex-brat. The only choice then was of staying as you were, waiting for someone to die or retire, or alternatively giving up either the air or ground part of your trade and moving up the one step to chief technician on a time of service basis, knowing that there was no slot available to go on to master technician. In other words, you could add a crown to your three stripes a little sooner, pocket a few shillings a day more, but you had to give up the idea of ever achieving warrant rank, and accept the very real risk of exposing yourself to the dangers of now only having a part of your original trade in a shrinking Air Force. As things turned out, that risk was great. Many of us believed as we left Halton that our extra technical competence alone could put us on a faster track in the service, but we were quite mistaken. What the service wanted was that we submit to the system, roll with the punches and let it carry us forward at its own pace, and that any higher level of competence was not really involved at all.

COMING HOME

He's my brother Silvest
Got a row of fifty medals on his chest

Forty-three years after our graduation, I returned for the first time to Halton. The occasion was the graduation of the One Hundred and Fifty-Fifth entry, and the last entry of all. I first learned of this impending event from my son, who is himself a member of the R.A.F. now, even if it does happen to be the wrong one. In my preliminary enquiries I also made contact again with some of my Fifty-Fifth entry comrades, including 'Slash', now living in Australia, and 'Horace', who seem to have featured quite often in these pages, as they did in my young apprentice's life. As the letters began to arrive through my foreign letter box, and I read of what had happened to them in life, I came to understand that this deep involvement is still far from over for many of us. We are seemingly all increasingly protective of each other, recognising that we are in fact a very much endangered species. The upshot was that I, along with 'Horace', Robin Berry and more than a handful of others from my entry were there on the day that the last apprentices passed out, and it is entirely in keeping that I tell something about my feelings returning to Halton after an absence of more than forty years. The still barely more than early middle-aged prince was re-entering the stage, if only as a day tripper. Probably the reader will be curious as to my reasons for going back at all, holding the views on the service which I apparently do, but I feel that I still owe so much to my former comrades, and particularly to those of us who were not so fortunate as to make it this far, and for reasons that my comrades of all entries will so readily understand and the Air Force will never truly comprehend at all, I am still proud to have been an apprentice.

The evening before the parade had been spent with 'Horace' as a guest at his home, first reminiscing then catching up on our forty years' separation until the wee hours, and sharing my duty-free and his drinks cabinet until neither was sure exactly who would push and who would ride in the wheelchair on the morrow. Life does seem to have its ways of turning full circle, and for all of us too. I heard from 'Horace' of the day that he, then commissioned and a flight lieutenant, met up with none other than our old friend the regiment sergeant, by then a unit warrant officer, and apparently a good one too, as 'Horace' chanced to pass the latter's married quarter garden as he was busily engaged in pruning a large bush. There was never any ongoing bitterness or rankling with 'Horace', he was never that sort of chap, and so it seems, neither was our old friend. After exchanging pleasantries and reminiscences, 'Horace' remarked that surely the warrant officer's life was now complete, having his very own bushy-topped tree to care for and they shared the joke together. We are, it seems, all so much more mellowed by time, and it has been quite difficult to try and write as I felt at that time, and not as I now feel about it all forty years on. With all its faults, the system still did manage to pass through its portals more than a fair share of people of true character and, who knows, perhaps some of that was due to people like our old friend, with whom we have not always shared deep mutual affection.

Came the agonising headache morn and the first point of non-agreement was reached as early as in the station car park, where my son introduced me to a former colleague of his who was, it seemed, none other than the officer responsible for curbing any One Hundred and Fifty-Fifth Entry, shall we say, vandalistic tendencies on this, their last day. I recognised the facial expression at once of course despite the long time interval. His eyes darted restlessly here and there, and he twitched nervously at any sudden loud or unexpected noise. Unaware of my origins, he confided in me as an older man and responsible parent, and I was told with horror that on a previous occasion an entry had even engraved their number in the sergeants' mess bowling green with weedkiller, and both my son and his friend agreed gravely with much head shaking that this was most surely, "Well over the top". My blank non-comprehending reply of "Over the top of what?" rather gave the game away about my true origins, I am afraid, and revealed an apparently ongoing philosophical difference of viewpoint.

I already knew of an entry junior to mine that had been fortunate enough to have had a wonderfully ripe situation delivered on their very doorstep when the forecourt of Wendover railway station had been in the process of renovation, which just happened to coincide with their passing out. They had diligently applied their so recently acquired skills to the construction of a rather better than passable imitation of a bomb and under cover of darkness, had nipped out of camp and buried it not too deeply beneath the recent and still ongoing excavations. When the workmen had climbed into the hole the following morning to renew their labours, it had not taken them too long to uncover this new interesting find, and they had apparently abandoned their tools and removed themselves to a safer distance as a matter of some pressing urgency. The end result was that the immediate area was cordoned off and rail traffic on the whole line had been totally discontinued until the bomb squad had been able to open the metal canister. There they had recovered a folded written message which imparted a perhaps succinct but nevertheless meaningful piece of information: "Fff you, courtesy of the NNNth entry."

That I consider was perhaps barely marginal, although I was personally frankly envious when I heard of it, and somewhat sad that railway renovations had not been available to my own entry as a viable option.

I did consider asking my son's young friend if he was perhaps so fortunate as to be accompanying the entry on their postings into the big world R.A.F., and if so if they would be travelling by train, but I decided to forgo that pleasure. If his superiors had seen fit to trust him with such a high responsibility as the station car park for the day then perhaps they knew him better than I, and he was at a very impressionable young age after all.

I of course took the opportunity to stop off on my way to the Henderson parade ground, and paid homage to number one wing guardroom with its memories of adjacent cell mates and Parents Day, now with its heavily wire-caged and bricked-up general facade, most obviously intended to deter people from getting in. In my day the internal architecture, with which I had once been all too familiar, was primarily intended to prevent one from getting out, although even then in Pat's case it had abysmally failed. I also even had time for a brief peep at Maitland Square, now in use as a car park, and almost heard the bellow of 'Steve' demanding that they remove their vehicles at once from HIS parade ground. The whole clock tower on the number three wing 'tank', which 'Rigor' Smith used to regularly climb at night in order to stop the clock, and on one occasion even crowned with a chamber pot, has somehow disappeared over the years. Maybe they eventually grew tired of having to regularly enlist the aid of the station fire department with its tall ladder in order to start the clock yet again.

At first, my impression of the whole occasion was that of a disappointing, vastly shrunken affair, which seemed a very much smaller sideshow of my own passing out, with only very small and barely recognisable coincident themes. No more apprentices' pipes and drums or military bands, which were at the very soul of our being, just strangers and in tarted-up fancy dress. They wouldn't even have merited a place on the same parade ground in my day. The impression was one of so much less. After all, thirty-odd apprentices on the parade ground simply cannot compare with the spectacle of the two thousand of my time, but, as the day progressed, I discovered a few small signs that perhaps all had not been entirely lost over those long intervening years.

A decorative floral patch in front of some headquarters building or other, which seemed to have a certain unusual pattern, aroused my curiosity, and on closer examination proved to have written in flowering bloom an expression of four-lettered vulgar sentiment quite remarkably similar to the Fifty-Fifth electricians' banner, courtesy of one of the One Hundred and Forty-something, but there the weeds had sadly encroached. The long-term planning involved in ensuring that this sentiment would come to light again every spring began to impress me no end. The parade square had of course been marked out in white paint for markers positions and flights, and enclosures for guests had been similarly indicated. I do not believe for one moment that it was the same personnel who had added by chance a sign which had a capital W followed by a drawing of an anchor plus a capital S, outside the senior officers' enclosure. A feeling of quiet pride and deep affinity with the perpetrators of this prank came over me, I freely confess.

Many of us ex-brats including all the Fifty-Fifth Entry attendees had been accommodated, wisely in my view, well back from the parade ground and safely away from any chance contact with the distinguished royal visitor, which I personally regarded as being a very fine compliment indeed and one showing quite a remarkable capacity for suspicious memory, what is more. Who knows what questions may have been asked in any light exchange of dialogue concerning the young, female and frequently-in-the-tabloid-press members of his illustrious family. Not all of the ex-brats around me saw it in quite that light, however. We had not been provided with the comfort of chairs or printed programmes for ourselves or our ladies either, which, in any case, on past form we would undoubtedly have vandalised at the earliest possible opportunity, given the chance. The chairs were a simple matter and we walked to the nearby number two wing 'tank' and got them for ourselves, watched by the grinning 'boggies' who had been detailed for, but of course failed to carry out, the job. When it was all over, we left the whole lot standing just where they were and smiled in turn at the same now crestfallen 'boggies' as we went on our merry way.

The matter of the programmes was to me just a simple initiative test being applied all over again after all those years, and so I moved myself in a smart ex-apprentice-like manner to an advantageous position close to the saluting base, and enquired of a very attractive young W.R.A.F. officer in attendance there where the senior officers' enclosure might be. She couldn't have been more helpful and charmingly led the way and even opened a white painted wicket gate in a splendid enclosure for me to enter. I did so, gathered up a half dozen or so programmes from the seats, winked in my best suggestive manner at her on my way out in response to her incredulous open-mouthed gaze, and brought them back to my comrades and their ladies at the back of the parade ground. As Robin Berry saw me returning with my shared offering he remarked loudly for all to hear, "Hello! 'Grim's' back," and indeed, just for the day, I was.

When the trumpet call for march on sounded, even though it was no longer played by an apprentice, just for a brief fleeting moment I could have been back in 1949 all over again, and raring to go. The old tingle returned, and so it did for most of us ex-brats watching, judging by the faces around me. Suddenly it wasn't just the last entry's day any more – it was equally, if not more so, all of ours.

When first the Red Arrows and then all the other flying units beat up the parade ground, it was the Air Force at long last acknowledging and paying its due to every single one of us who had ever stood there over the years, and in the way that we understood best. The volunteer supporting flights of ex-apprentices were made up of mixed ranks ranging from air commodore to junior technician, and they stood side by side in the ranks as proudly as they had on their very own special day, regardless of their present rank or seniority. There was one former member of the Fifty-Sixth entry, and even the goat handler was a Flight Lieutenant. For this one special day, they and the R.A.F. were above all that kind of thing. Probably the goat itself was at least an air vice marshal by that time, and accordingly it even had a younger aide-de-camp goat companion in attendance. Both left an impression on the parade ground, by the way. If there were to be no junior entries for the last entry to march off through, then there were to be their seniors to support them to the very last. It was a very emotional and touching experience for me, because, after all, I am very much a part of all that. It was also a privilege to share it all and be reunited with my old and dear friend 'Horace', and to push his wheelchair for the day, neither of us knowing that it was to be my first and last opportunity to do so.

The very last comment of all on undergoing the total apprentices' experience I must concede to my old friend, sadly no longer with us. As I pushed the wheelchair away from the parade ground on our way to the car park and home after the final passing out parade, we chanced to pass the 'boggie' supporting flight drill squad, now at ease but still under the command of their flight sergeant drill instructor. Recognising who we both probably were, and our likely professional eye for the just demonstrated parade skills of his charges he proudly enquired, "Did you enjoy it then?"

'Horace' and I looked at each other and smiled indulgently, then 'Horace' replied aptly for both of us, referring of course to the much earlier and larger event that had gone on for our three years at Halton, rather than the short spectacle that we had just witnessed.

"Well, some of it was quite funny now that you mention it, yes", he said.